北京建筑大学教材建设项目资助出版

简明建筑英语翻译教程

主 编 杜 苗

中国财经出版传媒集团
中国财政经济出版社

图书在版编目（CIP）数据

简明建筑英语翻译教程／杜苗主编．--北京：中国财政经济出版社，2019.12

ISBN 978-7-5095-9511-4

Ⅰ.①简… Ⅱ.①杜… Ⅲ.①建筑-英语-翻译-高等学校-教材 Ⅳ.①TU

中国版本图书馆 CIP 数据核字（2019）第 291202 号

责任编辑：彭　波　　　　　责任印制：党　辉
封面设计：卜建辰　　　　　责任校对：李　丽

中国财政经济出版社 出版

URL: http://www.cfeph.cn

E-mail: cfeph@cfemg.cn

（版权所有　翻印必究）

社址：北京市海淀区阜成路甲 28 号　邮政编码：100142
营销中心电话：010-88191537
北京财经印刷厂印装　各地新华书店经销
710×1000 毫米　16 开　26.25 印张　346 000 字
2019 年 12 月第 1 版　2019 年 12 月北京第 1 次印刷
定价：68.00 元
ISBN 978-7-5095-9511-4
（图书出现印装问题，本社负责调换）
本社质量投诉电话：010-88190744
打击盗版举报热线：010-88191661　QQ：2242791300

主　审　李宜兰
副主编　王　隽　刘晓玉
编　者　杜　苗　李宜兰　王　隽
　　　　　　刘晓玉　王鲁豫　王彩霞

目　　录

第一部分　翻译的基础知识 ………………………………… 1

 1.1　翻译的定义 ……………………………………… 3
 1.2　翻译的目的 ……………………………………… 4
 1.3　翻译的标准 ……………………………………… 5
 1.4　翻译的单位 ……………………………………… 10
 1.5　翻译的分类 ……………………………………… 15
 1.6　翻译的过程 ……………………………………… 16
 1.7　译者的素质要求 ………………………………… 23

第二部分　翻译中的英汉语言对比 …………………………… 27

 2.1　词汇 ……………………………………………… 29
 2.2　语法（句法结构） ……………………………… 36
 2.3　语序 ……………………………………………… 41
 2.4　修辞 ……………………………………………… 48

第三部分　翻译中非语言层面的对比 ………………………… 63

 3.1　思维 ……………………………………………… 65
 3.2　文化 ……………………………………………… 77
 3.3　习俗 ……………………………………………… 80
 3.4　历史 ……………………………………………… 84

3.5　社会心理 ……………………………………………… 87
　　3.6　审美 …………………………………………………… 91

第四部分　翻译技巧及建筑文本翻译训练 ………………… 97
　　4.1　翻译技巧：增益法和删减/省略法 ………………… 99
　　4.2　翻译技巧：重复法和正反译法 …………………… 131
　　4.3　翻译技巧：状语从句的译法 ……………………… 161
　　4.4　翻译技巧：定语从句的译法 ……………………… 194
　　4.5　翻译技巧：名词性从句的译法 …………………… 231

第五部分　建筑文本翻译拓展训练（一）英译汉
　　　　　——Gothic, Rococo and Mediterranean Architecture …… 265
　　5.1　文本原文 …………………………………………… 267
　　5.2　Terms and Vocabularies …………………………… 288
　　5.3　Syntactic Structure 翻译及重点句型分析 ………… 292
　　5.4　参考译文 …………………………………………… 299

第六部分　建筑文本翻译拓展训练（二）汉译英
　　　　　——浪漫主义及功能主义建筑 …………………… 315
　　6.1　文本原文 …………………………………………… 317
　　6.2　参考译文 …………………………………………… 332
　　6.3　Vocabulary 词汇表 ………………………………… 351
　　6.4　Syntactic Structure 重点句型分析 ………………… 359

第七部分　建筑学常用专业术语汉英对照 ……………… 365

参考书目 ……………………………………………………… 403

简明建筑英语
翻译教程
Chapter 1

第一部分　翻译的基础知识

1.1 翻译的定义

翻译是一种语言活动,人们进行语言活动主要就是为了交流思想。把一种语言所表达的内容用另外一种语言表达出来,这一过程即为翻译。从社会语言学的观点来看,语言是脱离不了社会文化的,使用不同语言的人,往往处于不同的文化背景。因此语际之间的交流实际上是在进行不同文化间的交流。

一些知名的学者和翻译家对"翻译"给出如下的定义:

张培基认为翻译是运用一种语言把另一种语言所表达的思想内容准确而完整地重新表达出的语言活动。这里强调翻译是一种语言活动,用以准确且完整地传递信息内容。

我国翻译家黄龙在其英文版著作《翻译学》一书中将翻译定义为:"Translation is the unity of opposites wherein an equivalent and aesthetic intercommunication of bilateral alien languages (letter language, semiotic language, animal language) in social sciences (including theology in a broad sense) and physical sciences is performed theoretically and practically through oral interpretation and written translation by the agency of human brain or electronic brain." 黄龙的定义非常全面,即翻译不仅是双语之间的信息转换,还应当考虑审美方面的要求;翻译不但涵盖社会科学等众多学科,而且涉及自然科学的诸多门类;翻译既有理论也有实践;既可以是口译或笔译,又可以是机器翻译。

美国翻译家和理论家奈达(Eugene A. Nida)在《翻译理论与实践》一书中给"翻译"下了这样的定义:所谓翻译,是指在译语中用最切近而又自然的对等语再现原语信息,首先在语义上,其次是文体上。(Translation consists in reproducing in the receptor language the closest natural equivalent of the source – language message, first in terms of meaning and secondly in terms style）。在这里奈达强调了四点：①再现原语信息（reproduction）；②原语与译语（source language vs. receptor language）；③对等语（equivalent）；④语义与文体（meaning vs. style）。

奈达认为翻译的功能在于"再现原语信息"，即将原语信息转换为译语信息，"信息"的转换不仅表现在"语义"上，而且表现在"文体"上；奈达的定义也描述了原文与译文之间的关系：译文一方面要取得与原文"最切近"的对等效果，另一方面必须是"地道的"译语语言。

卡特福德（J. C. Catford）认为翻译是把"一种语言（译出语）的话语材料转换成另一种语言（译入语）中对等的话语材料"（the replacement of textual material in one language by equivalent textual material in another language）。

与翻译的语言学派关注语言结构和语言形式对应问题不同，翻译的文化学派将注意力转向翻译在文化传递方面的重要作用。他们认为翻译绝非一种纯语言的语际转换活动，而是一种文化间的交流活动。巴奈斯特（Bassnet）认为"翻译就是文化内部与文化之间的交流"。翻译不应局限于语言上的对等，而是在译语文化中实现与原文文化功能的对等。为此，翻译应突破以语篇为单位的传统观点，以文化为单位，重新审视原语文本和译文文本在各自文化系统中的功能和意义，以及译者的地位、翻译与文化、亚文化之间的关系等问题。

综上所述，翻译包含两层含义：第一，翻译是一种语际转换活动，其转换的主要内容为原语信息；第二，从原语到译语的信息转换不仅是内容上的，同时也应考虑到原作的文体风格和文化功能。

1.2　翻译的目的

从宏观上看，翻译作为一种语际转换活动，涉及社会生活的方方面面。世界上有近3000种语言，其中使用范围较广的语言有10几种，这就给使用不同语言的人进行信息交流带来了很大不便，而翻译就是一座沟通不同语言使用者之间的桥梁。可以说，没有翻译，这个世界就是支离破碎的，就是孤立的、封闭的、相互隔离的。纵观我国的历史，曾出现过三次翻译的高潮：东汉至唐宋的佛经翻译、明末清初的科技翻译和鸦片战争至"五四"运动的西学翻译。每一次的翻译高潮无不伴随着文化交流和思想的解放运动。翻译在国际政治、经济、军事、科学、文化、艺术等各个领域进行不同语言之间的信息转换与传递，促进了这些领域世界性或区域性的交流，为世界文明的发展与和谐做出了巨大的贡献。

从微观上讲，翻译作为外语本科阶段的一门必修课程，其重要性不言而喻。在学习中的听、说、读、写、译五个技能中，译是最后一个，也是最难和最高级的一个技能。翻译技能的提升能促进其他技能的熟练掌握与运用。

随着全球交流的越发频繁和便捷，越来越多的人将会面临国际交流，将来不可避免地要承担起文化交流的重任。借鉴、吸收、引进外国先进的知识、技术和经验，并将我国的优秀文化介绍给外国都需要语际之间的交流翻译。学习和掌握翻译技能、提高翻译能力是实现这些目标的必要条件。

1.3 翻译的标准

翻译的标准是衡量翻译的尺度。有了好的标准，翻译活动才会有努力的方向，才会保证翻译的质量。关于翻译的标准，我国历来就有众多的看法。在汉唐时期，便有"文质"之争。主张"文"的翻译家强调翻译的修辞和通顺，强调翻译的可读性。主张"质"的翻译家则强调翻译的不增不减，即翻译的忠实性。"文质之争"实际上是意译与直译之争。

清代著名翻译家严复一生翻译了大量的西方政治经济著作,他根据自己的翻译实践于1898年在《天演论》卷首的《译例言》中提出:

"译事三难:信、达、雅,其求信已大难矣!顾信矣不达,虽译犹不译也,则达尚焉。……译文取明深义,故词句之间,时有所傎倒附益,不斤斤于字比句次,而意义则不倍本文。

"假令仿此(西文句法)为译,则恐必不可通,而删削取径,又恐意义有漏。此在译者将全文神理,融会于心,则笔下抒词,自善互备,至原文词理本深,难于共喻,则当前后引衬,以显其意。凡此经营,皆以为达;为达即以为信也。

"易曰:'修辞立诚。'子曰:'辞达而已。'又曰:'言之无文行之不远。'三者乃文章正轨,亦即为译事楷模。故信、达而外,求其尔雅。"

这就是我国奉为臬圭的"信(faithfulness)、达(expressiveness)、雅(elegance)"三字标准,这个标准对后世影响极大。"信"指译文须忠实于原文,"达"指译文通达。严复认为"信"是排在首位的,然"顾信矣不达,虽译犹不译",不"达"在某种意义上也就是不"信","信"与"达"的关系是辩证的,是不可分割的两个方面;关于"雅"字,争论较多——有人认为"雅"指的是雅正,严复把当时的士大夫作为译文的阅读对象,使译文适合读者的接受水平,有利于译文的传播,从这个意义上看,"雅"的原则至今仍具有积极的意义,与读者接受理论有异曲同工之处。

鲁迅在讨论翻译标准时说:"凡是翻译,必须兼顾两面:一当然力求其易解,一则保存着原作的丰姿",但不该"削鼻剜眼"来达到顺眼的目的。针对赵景深"与其信而不顺,不如顺而不信"的观点,鲁迅阐述了自己"宁信而不顺"的直译观点:"自然,这所谓'不顺',决不是说'跪下'要译作'跪在膝之上','天河'要译作

'牛奶路'的意思,乃是说,不妨像吃茶淘饭一样几口可以咽完,却必须费牙来嚼一嚼。"这里可以看出鲁迅主张的直译并非是"硬译"和"死译"。

在《论翻译》一文中,林语堂提出"翻译的标准问题大概包括三方面。第一是忠实标准,第二是通顺标准,第三是美的标准",并将这三条标准概括为译者的三种责任:"第一是译者对原著的责任,第二是译者对中国读者的责任,第三是译者对艺术的责任。三样的责任心备,然后可以谓具有真正译家的资格。"林语堂把翻译看作是一门艺术,他也是中国译学史上第一个最明确地将现代语言学和心理学作为翻译理论的"学理剖析"的基础的人,因此他的"忠实、通顺、美"的翻译标准是以语言学、心理学和美学为依据的。他把翻译标准与译者的责任联系起来,独树一帜。

1951年,傅雷提出文学翻译应当"神似"的观点。傅雷认为,"以效果而论,翻译应当像临画一样,所求的不在形似而在神似"。这种"神似"论是建立在东西方人在思维和表达方面的差异基础上的,"传神云云,谈何容易!年岁经验愈增,对原作体会愈增,而传神愈感不足。领悟为一事,用中文表达为又一事。东方人与西欧国家人之思想方式有基本分歧,我人重综合,重归纳,重暗示,重含蓄;西方人则重分析,细微曲折,挖掘唯恐不尽,描写唯恐不周:此两种mentalite(心智类型)殊难彼此融洽交流。……愚对译事看法实甚简单,重神似不重形似。"

1964年,钱钟书提出了翻译的"化境"论:"文学翻译的最高标准是'化'。把作品从一国文字转变成另一国文字,既能不因语文习惯的差异而露出生硬牵强的痕迹,又能完全保存原有的风味,那就算得入于'化境'。17世纪有人赞美这种造诣的翻译,比为原作的'投胎转世'(the transmigration of souls),躯壳换了一个,而精神姿致依然故我。换句话说,译本对原作应该忠实得以至于读起来不像译本,因为作品在原文里决不会读起来像经过翻译似的。"但是,正如钱钟书自己所说的:"彻底和全部的'化',是不可能实现的理想。"

在西方,对于翻译的标准也有不同的说法。早在1789年,英国翻译家、学者乔治·坎贝尔(George Campbell)就提出了翻译的"三原则",即译者必须:第一,准确地再现原作的意思;第二,在符合译作语言特征的前提下,尽可能地移植作者的精神和风格;第三,也是最后,使译作至少具有原创作品的特征,显得自然流畅。(The first thing…is to give a just representation of the sense of the original…The second thing is, to convey into his version, as much as possible, in a consistency with the genius of the language which he writes, the author's spirit and manner…The third and last thing is, to take care, the version have at least, so far the quality of an original performance, as to appear natural and easy…)

1792年,英国著名学者亚历山大·弗雷赛·泰特勒(Alexander F. Tytler)也提出了一个"三原则":第一,译作应完全复写出原作的思想;第二,译作的风格和行文手法应与原文保持一致;第三,译作应具备原作的流畅感。(That the translation should give a complete transcript of the ideas of the original work; That the style and manner of writing should be of the same character with that of the original; That the translation should have all the ease of original composition)。

奈达以《圣经》翻译实践为基础,提出了功能对等的"等效理论"。最初奈达的"等效理论"使用动态对等(dynamic equivalence)这一术语,后为强调"功能"这一概念改用"功能对等(functional equivalence)"这一术语。在讨论功能对等时,他把语言看成一种交际形式,翻译则是一种交际活动,一种介于不同语言与文化之间的交际活动。交际的目的是使参与交际的各方能够相互理解与沟通。因此,翻译首先就是要"再现原文信息",使原文的信息能够在译语的文化背景和交际场景中取得与原语同样的交际效果,从而保障交际的顺利进行,使交际各方进行有效地沟通与交流。在奈达看来,内容是第一位的，形式是第二位的；但他也认为，功能对等的翻译，不仅包括信息内容，同样也包括语言形式，即"在译语中用最贴近而又自然的对等语再现原语的信息，首先在语义上，其次是文体上"。

另外，还有雅各布森（Roman Jakobson）、卡特福德的等值论，弗美尔（H. Vemeer）的目的论，以及后现代主义、解构主义，女性主义等理论都对翻译标准提出了自己的标准和看法。

不过，对于初学者而言，"忠实、通顺"是做好翻译工作的最基本标准，只有在此基础上，才能一步步地向翻译的更高标准迈进。所谓忠实，首先要忠实于原作的内容。译者必须把原作的内容完整而准确地表达出来，不能随意地篡改、歪曲、增加和删除原作的内容，这里的原作内容通常指的是作品中所叙述的事实、说明的事理、描写的景物以及作者在叙述、说明和描写过程中所反映的思想、观点、立场和所流露的感情等。忠实还包括保持原作的风格，即原作的民族风格、时代风格、语体风格、作者个人的语言风格等。译者对原作的风格不能任意破坏和改变，不能以译者的风格来取代作者的风格。原作如果是通俗的口语体，则不能译成文绉绉的书面体；原作如果是粗俗琐屑的，则不能译成文雅洗练的；原作如果是富于西方色彩的，则不能译成富于东方色彩的。总之，原作如何，译文也应该如何，尽可能还原其本来面目。所谓通顺，即指译文语言必须通顺易懂，符合规范。译文必须是明白晓畅的现代语言，没有逐词死译、硬译的现象，没有语言晦涩、佶屈聱牙的现象，也没有文理不通、结构混乱、逻辑不清的现象。当然，译文的通顺程度还应当要与原文的通顺程度相一致。例如，在文艺作品中，作者有时会有意识地采用或引用一些非规范语言以刻画人物或渲染某种气氛，翻译时应设法如实传达，不应加以改变。

忠实与通顺是密不可分、相互依存的两个方面。忠实而不通顺，读者看不懂，或者看不下去，忠实便失去了意义；通顺而不忠实，脱离了原作的思想内容与风格，成了"胡译"和"乱译"。可以说翻译应力求"信与顺",力戒"宁信而不顺"或"宁顺而不信"。

1.4 翻译的单位

首次提出"翻译单位"这个概念的 Vinay 和 Darbelnet 将翻译单位定义为"思维单位";它在语言上表现为"最小的、无需分开翻译的语言单位";苏联翻译理论家巴尔胡达罗夫在《语言与翻译》一书中把翻译单位定义为"在译文中能够找到对应物的原文单位,但它的组成部分单独地在译文中并没有对应物。换言之,翻译单位就是原语在译语中具备对应物的最小(最低限度)的语言单位"。翻译的单位指的是翻译过程中的操作单位,同时也是语言转换发生的层次。划分翻译单位的目的在于对语言进行科学有效的语法和语义分析,以便选择相应的翻译方法。因此翻译单位也就是用以分析的有意义的语言单位。下面简要介绍英汉两种语言在各个语言单位层次上的转换,并结合翻译单位的属性特征,提出在翻译实践中可操作性较强的翻译基本单位。

1.4.1 英汉语言单位与翻译

汉语和英语的语言单位可以概括为音位(字位)、词素、词、词组(短语)、句子、段落、篇章这七个层次。两种语言的对等转换、不对等转换、跨层次转换主要集中在这些层面上。

1. 音位(字位)层(phoneme – level)翻译

音位本身没有独立的意义,但在音译时,要求原语的音位在译语中找到对应物,音位就成为翻译单位,自成一体,如人名、部分地名以及一些通用的国际计量单位的翻译等。如:Tom(汤姆)、Singapore(新加坡)、北京(Beijing)、ohm(欧姆,指电的单位)、volt

(伏特，指电压单位)、ampere（安培，指电流强度单位）、logic（逻辑）、humor（幽默）、sonar（声呐），另外还有 talk show（脱口秀）、mosaic（马赛克）。

音译作为翻译外来语的常用方法之一，需要遵守约定俗成或公认的标准，以求统一，同时也应当本着使译名通俗易懂、明白晓畅的原则。有些音译名读起来绕口别扭，在使用当中逐渐被淘汰，失去了生存的空间，例如："电话"（telephone）替代了"德律风"，"最后通牒"（ultimatum）替代了"哀的美敦书"，"灵感"（inspiration）取代了"烟士披里纯"一词。取而代之的新词生动形象，更易于思想意义的传递。

2. 词素层（morpheme - level）翻译

词素是语言中最小的有意义的单位。相对于音位层，词素层翻译出现的频率较低，但在科技和医药生化英语中，却有着广泛的应用。如：

electro -	magnet -	ic
电	磁	的
morpho -	gene -	tic
形态	基因	的
multi -	media	
多	媒体	
hi -	tech (technology)	
高	技术	

另外，常用的词素还有 psycho -（精神的心理的）psychotherapy（心理疗法）、psychology（心理学）、psychologist（心理学家）、psychological（心理学的）；hydr（o）-（水的、含氢的、氢化的）：hydrocarbon（碳氢化合物）、hydroelectric（水电的）、hydrogen（氢气）、hydroponics（水培）、hydrotherapy（水疗法）。英语具有庞大的构词系统，而词素在其中起着核心的作用，掌握这些词素及其用法，

对于译者来说是非常有益的。

3. 词语（单词）（word）层翻译

词是能独立运用的最小语言单位。在两种语法和句法结构较为相似的语言中，把词作为翻译单位是很常见且易于操作的。英汉两种语言的句法结构有一定的相似性。

4. 词组（短语）层（phrase – level）翻译

词组是比词大一级的单位，词对词的翻译容易导致死译、硬译，因而需要在更大的语境中寻求语义的等值。词组不是单个词的简单叠加，其意义是固定的、约定俗成的。

5. 句子层（sentence – level）翻译

在词组这一层次进行转换也不能正确理解原语的情况下，就需要以句子为单位来考察原语了。

6. 段落层（paragraph – level）翻译

段落是大于句子的语言单位，它可能是几个句群，也可能是一个句群，还可能只是一个句子、甚至是一个词。以段落为单位能够较充分地考虑到句与句之间的逻辑关系和语义关联，在翻译时对段落进行重新组织，使译文符合译语的表达习惯。汉语重意合，而英语重形合。

7. 篇章层（text – level）翻译

篇章是比句子、段落更大的语言单位，即由段落组成的、结构和语义完整的交际单位。以篇章为单位来考虑翻译可以对文本意义有更为完整的了解：原文的主题意义是什么，作者有什么样的思想观点，进行人物的形象分析、个性分析以及意境营造、情节发展分析等，以便在表达时恰当地运用翻译策略与方法；同时也可以对篇章类型，文体风格有更为全面的把握，以减少误译。这也就是所谓的"整体效应"，即从大处着眼，小处着手。

1.4.2 翻译的基本单位

从"语言单位"（音位、词素、词、词组、句子、段落、篇章等）来看，词是语言的基本单位。因为人类的语言是词的语言，它既能充当词汇意义与语法意义的载体，又能担任句子的任何成分，甚至可以独立成句。在实际的交际过程中，词还能完成语用功能。在作文或文学创作中，词的选择、推敲对表意、传情具有非常重要的意义。

既然词是语言的基本单位，那么它也应该成为翻译的基本单位。有学者认为词是思维的支撑。如果将词（包括词组）作为翻译的基本单位，它也符合最先提出"翻译单位"这个概念的两位学者的看法，因为词（词组）在翻译中属于不可再分割的有意义的单位。从语言学和翻译实践角度看，词和词组是句子的组成部分，是译者换码时首先予以注意并转换的一个成分。英国语言学家 J. R. Firth 说："每一个词在不同的上下文里都是一个新词"。尽管他强调的是上下文对单词意义的影响，但同时也揭示了词及其词义的重要性。对翻译来说，词（包括词组）是可操作的、有意义的、不可再分割的一个单位，是翻译过程中必须仔细分析对比、必须首先理解与转换的一个基本单位。纽马克认为："翻译必须以词语、句子和语义为基础。因为除此之外，翻译便是无本之木。没有词语，就不存在意义"。比如翻译"Darrow had whispered throwing a reassuring arm round my shoulder"这句英文，我们首先将原文切分成三个部分 Darrow//had whispered//throwing a reassuring arm round my shoulder；然后再将第二部分切分为 throwing//reassuring arm//round//my shoulder；最后再分析 a reassuring arm 的真正含义。全句可译为"达罗搂着我的肩膀，悄悄地对我说了此话，叫我不要担心"。

如 James Joyce 的"*Finnigan's Wake*"一书书名的汉译，也很能说明问题：最早的译文是《芬尼根人的觉醒》，给读者的印象是"芬尼根开始革命了"，后来正式译名变为《为芬尼根人守灵》。两个译文悬殊这么大，原因是第一个译者没有充分理解英文 wake 这个词在原文句中的含义，属于"望文生义"。正式译名是根据小说内容来理解"wake"的真正含义而翻译并确定的。

从教学实践看，翻译教学也总是从词义的确定和翻译讲起，从翻译方法上讲，所谓的"转换法""省略法""增词法"也大都从词或词组翻译的角度出发的。所以，词（词组）既是语言的基本单位，也是翻译的基本单位。

1.4.3 翻译单位的特征

如前所述，翻译单位必须是有意义的语言单位，此其一。其二，翻译单位也应该是辩证的、相对的；动态的、变化的；多元的、多层次的；描写性的、可操作性的，能对翻译实践起指导作用的。下面我们简要地谈一谈翻译单位的属性特征。

1. 相对性

如果认真研究一下上述种种所谓的翻译单位，我们就会发现：无论哪种分法，翻译单位都是有大有小，大中有小。此如段落，段落通常由句子组成，但句子（尤其是一个长句）也可成为一个段落。由此可见，所谓翻译单位是辩证的，大小是相对而非绝对。翻译实践也告诉我们：翻译中没有，也不大可能有个能适用于各类文体（文学、科技、政治、经济、实用）、各种翻译类型（笔译、口译、机器翻译）和不同翻译者（一般译者、职业翻译者和翻译家）的所谓的"翻译单位"。

2. 动态性

翻译活动是一种语际间、文化间的交流活动，翻译者通过大脑复杂的搜寻、理解、对比、分析、综合等活动，使源语言与目标语言实现交流。这种交流过程是动态的、富有变化的，而绝不是孤立的、静态的、一成不变的，更不是机械地用所谓的"翻译单位"去处理原文，按照词典释义套译或硬译原文中的词语和句子。从这个意义上说，翻译单位在实际翻译过程中，必须是弹性的，可大可小，翻译者也要根据翻译需要可随时缩小和放大翻译单位。

3. 多元性

翻译过程是一个不断对比分析、选择与综合的动态过程，译者在这个过程中扮演的角色是积极的、能动的。为了从内容和形式上再现原文信息，译者必定会调动一切有利的手段和方法。无论是词、词组、句子，还是段落、语篇；无论是结构段自然段，还是语义模块；无论是词素还是音素，译者都会紧紧抓住，予以重视。不同的视野，不同的方法，不同的研究人员对翻译单位的不同理解、不同的切分，其实也从另一个侧面说明翻译单位是多元的、多层次的。

4. 描写性

翻译单位多元、动态、相对的性质决定了它必须是描写性的而非规约性的。不管研究的角度如何，研究的方法如何，研究结论如何，对翻译单位的研究必须始终面向翻译实践，为翻译实践服务。

1.5 翻译的分类

翻译可从不同的角度进行分类。

从涉及的语言符号来分类，翻译可分为语内翻译（intralingual translation）和语际翻译（interlingual translation）。语内翻译指的是同一种语言文字内部不同方言（variety）之间进行的翻译。例如，中国普通话与各地方言之间的翻译；书面语体与口语体之间的翻译，如"五四"运动之前的文言文与白话文之间的翻译；或在不同文体之间进行的翻译，如将散文译成诗歌。语际翻译指的是不同语言之间所进行的语言转换活动，如将英语译成中文或将汉语译成英文等。一般情况下的翻译指的就是语际翻译。

从翻译的手段来分类，翻译可分为口译（interpretation 或 oral translation）、笔译（translation 或 written translation）和机器翻译（machine translation）；而口译又可以分为同声传译（simultaneous interpretation）、交替传译（consecutive interpretation）、视译（sight interpretation）等。

从翻译的题材来分类，翻译可分为应用文翻译（practical writing translation）、科技文献翻译（translation for science and technology）、文学翻译（literature translation）和一般性翻译（translation for general purposes）。应用文体主要包括各类公告、公函、启事、合同、通知等。一般来说，应用文体属于要式文本，格式规范、用语准确。科技文献主要指的是科技论著、科学资料、产品说明等具有专业特色的材料。文学翻译主要指的是文学和文艺作品，如小说、散文、诗歌、戏剧等，文学翻译常以美学为取向，追求译品的艺术等值。

从翻译的处理方式来分类，翻译可分为全译（complete translation）、摘译（partial translation）、编译（translating and editing）、节译（abridged translation 或 selective translation）、译述（interpretation）、综译（comprehensive translation）和译写（translating and writing）等。全译是指不加删节地将原文翻译出来。摘译是指译者根据具体的需要，选取原文的部分内容或章节进行翻译，一般仅摘取原文的核心部分或内容概要。编译是指译者对原文的内容进行编辑加工。节译指的是在翻译时允许译者在保持内容整体完整性的前提下，对原文进行部分地删节。译述指的是译者在对原文内容进行翻译时，加入了客观的介绍，以及自己的看法，而不拘泥于原文的语言表达。综译即综合性的文献翻译，是对同一专题的不同文献（包括不同语言的文本），通过节译和编译，作综合性的加工处理，全身一种符合特定需要的综合性译文文本。译写是指译者在翻译过程中融入自己的创作、想象和发挥，一部分是忠实原文，而大部分是译者自己的创作。

1.6 翻译的过程

翻译的过程可以分为三个部分：正确理解、恰当表达和仔细校正。

1.6.1 正确理解

理解是表达的前提,没有对原文准确透彻的理解,就不可能有恰当的表达。所谓"正确理解",就是要理解原文的词、短语、句子结构、修辞关系以及作者的写作意图等。一般说来,拿到一篇文章,应该通读几遍,结合上下文的词与词、词与句子、句子与句子、段落与段落以及整个文章之间的语法关系与逻辑关系,准确地理解作者所要表达的思想。首先应当了解的是:在任何一种语言中一词多义现象极为常见,这给词义理解带来了困难。英国著名语义学家杰弗里·利奇(Geoffrey Lech)在其著作《语义学》(Semantics)中将"意义"划分为七种类型,即概念意义(conceptual meaning)、内涵意义(connotative meaning)、社会意义(social meaning/stylistic meaning)、感情意义(affective meaning)、反映意义(reflective/reflected meaning)、搭配意义(collocative meaning)及主题意义(thematic meaning),并将其中的第二种至第六种意义用"联想意义(associative meaning)"来概括。

概念意义也称为指称意义或字面意义,它是词语意义的核心部分。由于文化差异,不同民族的人对于词义概念的界定存在着一定差异,如英语中的"wine"指的是"葡萄酒",而中文的"酒"则是一种酒精含量高的饮品。但有时某些英语学习者却在"wine"和"酒"之间建立一种直接、对等的联系,以至于一看到"wine"就直接翻译成"酒"。

内涵意义指的是词语概念意义之外的意义,是概念意义的属性,带有词语使用者的主观态度和感情色彩,时常难以区别。辨别内涵意义无论对于作者还是译者都尤为重要,因为内涵意义常常是区分近义词的关键,如 cry,shout 以及 scream 都可以表达"喊"的意思,但是它们的内涵意义是有很大区别的："cry"强调大声说、大声喊叫，较为常用；"shout"则是生气或想引起别人注意时的"喊叫"；"scream"用以形容因极度痛苦、恐惧或激动而发出的尖声叫喊。同样用以描述偷窃，"steal"指的是偷窃且不使用暴力，"rob"指的是暴力抢劫，而"burgle"强调强行进入室内盗窃。

社会意义或风格意义指的是词语在发展过程中带有的社会信息，如社会价值观念等，例如：在女权运动影响下产生的用以替代"chairman"的"chairperson"就明显地受到女权主义的影响；"negro"则反映出黑人一度受到强烈的种族歧视与隔离。

情感意义指的是词语直接或间接表达交际者的感情、评价、态度的褒贬等，如在英语中"politician"和"statesman"所带的评价是不同的；又如，"顽强"与"顽固"都是形容人做事不肯轻易放弃，却带有不同的感情色彩；再如"ambition"一词可根据上下文分别译为"野心"和"雄心"。

反映意义指的是通过与同一词语的另一意义的联想来传递的意义，也即是常说的联想意义（associative meaning）。如英语中的"owl"（猫头鹰）除了表示"做夜工、熬夜的人"这个意思外，常令人联想到"精明、智慧"和"神情严肃的人"之意，例如：An early owl called, but to Charles it seemed an afternoon singularly without wisdom（一只早早出巢的猫头鹰叫了起来，但对查尔斯来说，这个下午却过得毫无智慧可言），He is always as grave as an owl（他老是板起面孔，神情严肃）。汉语里，"猫头鹰"俗称"夜猫子"，也可以用来比喻夜里工作、晚睡的人。另外，因"猫头鹰"在深夜常发出凄厉的叫声，故在汉语里常常与"倒霉、厄运、不吉利"联系在一起，并被认为是一种不吉利的鸟，落在谁家，谁家就要遭殃。因此在汉语中有"夜猫子进宅，好事不来"等说法。

搭配意义指的是一个词与其他词搭配在一起所产生的意义。例如，看报纸（read newspaper），看病（see a doctor 或 see a patient），看不起（despise, disdain）、看成（regard as, consider as）、看穿（see through）、看得起（think highly of）、看重（value）、看来（it seems, it appears）、看见（see, catch sight of）、看法（view, opinion）等。

主题意义指的是信息的组织方式，如强调、语序、语调等所传递的意义，如 It is from the sun that we get light and heat（正是从太阳那儿我们获得了光和热），而 We get light and heat from the sun（我们从太阳获得光和热）则没有前一句的强调含义。

这也就从一个侧面看出在理解过程中确定词义的难度很大。到底应该如何做到准确地理解原文呢？从翻译实践的步骤来看，准确地理解原文首先要理解原文词语和句子的意思。从上文的意义分类可以得出，在任何一种语言中一词多义现象是极为常见的，想要弄懂原文意思，必须看词语在具体的上下文中所表达的具体意思。

首先，阅读原文时，要理清原文的逻辑关系（logical relationship）以便根据目标语言的习惯对译文进行相应的调整，使译文适应译入语的表达习惯。语篇内部意义上的联系，包括：顺承关系、因果关系、转折关系等逻辑关系。译者在分析原文结构时，应先找出原文句子的主干部分和重心所在，再分析句子的修饰成分。

其次，指代关系（reference）也是研究上下文时的一个非常重要的环节，如果对原文指代关系把握不准，很容易导致译文指代不明。表达指代系的方式主要包括指示代词、人称代词等，如 it、he、she、this、that、these、those、so 以及它、他、她、这个、那个等。

再者，在研究上下文时，除了应当仔细分析原文的逻辑关系、指代关系之外，要准确地理解原文的意思，必须利用背景知识，正确判断作者的写作意图，从而结合文章的思路选择适当的翻译策略和方法。

翻译是一种跨文化交际，翻译的原文中充满了富有文化内涵的词语，这些文化内涵构成了理解原文的障碍。因此，在理解原文的过程中,译者还必须考虑到词语的文化因素,以便正确理解原文的意思。

可以毫不夸张地说,没有准确的理解就没有翻译;翻译是架在不同文化之间的一座桥梁,译者要跨越时间、空间的距离,与原作者达到心神相通。译者不但要明了词语的确切意义和感情色彩,还要明了句子内外逻辑关系和整个语篇的意向所指。在理解原文这一阶段,遇到有疑问的地方或与上下文意思有出入的地方,务必要及时地查阅资料和字典,吃透原文,为表达阶段做准备。

1.6.2 恰当表达

如果说理解是进入原文的话,那么表达就是要跳出原文。对于原文的正确理解并不能确保表达得恰当。表达能否做到恰当也取决于译者对译入语的掌握与修养。翻译初学者往往受母语思维方式、语法结构和表达方式的影响,在表达方面常常出现"辞不达意"的情况,正所谓"理解难,表达更难"。另外,表达的恰当与否也与译者所采取的翻译方法有关。下面简要介绍一下直译与意译两种翻译途径。

1. 直译

直译(literal translation)指翻译时传达原文内容的同时尽量保持原作的语言形式,包括用词、句子结构、修辞手段等。在不影响读者对译文内容理解的前提下,直译可以保留原文传达的形象、比喻和地方民族色彩。

直译是在译出语的表达方式与译入语相同或相似的情况下采用的一种方式,直译后的译文与原文表达同样的效果,与逐词对译的死译不同。

作为初学者要警惕的是另一类死译,这类翻译晦涩难懂,不符合译入语的表达习惯,让人读后不知所云,如坠雾里。

2. 意译

意译(liberal translation 或 free translation)是指从意义出发,用

词、句法结构、修辞等手段将原文意义表达出来而不拘泥于原作的一种方式。意译可以使得译文更加向译入语靠拢，比起直译更容易为读者理解。意译能够根据上下文灵活地进行选词，按原文的逻辑关系对原文的句序进行适当的调整，使表达更加符合译入语的习惯。

翻译初学者要注意的是切不可将意译与天马行空般的"胡译""乱译"混为一谈。意译主要是在文化差异较大的情况下，为了更好地将意义传达给读者而采取的一种翻译方法，但意译仍然以"忠实"地传递原文内容为原则。"胡译""乱译"则将"忠实"的翻译标准抛于脑后，片面追求译文的"通顺"，望文生义，对原文擅自变更，导致译出来的东西与原文相去甚远、不堪卒读。

由上可见，直译以语言的共性和文化共性为基础，意译则基于语言的差异性与文化差异。也就是说，在思想内容一致的前提下，译文语言形式与原文语言形式完全对等的情况是存在的，在这种情况下要采取直译；如果原文思想内容的保留不得不以调整或改变原文的语言形式为代价，就得采用意译。直译的缺点在于：在译文与原文语言形式不完全对等的情况下，采用直译会使译文读起来更吃力；而如果在原文形式体现了意义的情况下，也即形式附带意义时，采用意译会失去形式附带的意义，如将汉语诗歌以散文的形式传译成英语就会损失汉语诗歌音韵美，因为平仄声调所带来的韵律美是汉语诗歌不可分割的特征。在具体的翻译实践中，直译与意译是交叉使用的，其运用是否得当必须以"忠实、通顺"为标准，好的译文应该准确地表达原文的意义，且遵循译语的表达习惯。一般说来，直译与意译在不同类型的篇章中运用的比例也是不同的，如法律、政治、新闻文本就要求更多字面对应的准确性较强的直译，而诗歌、散文等文学作品、广告语篇等则对意译情有独钟，以求更强的可读性。因此，采用直译还是意译必须考虑到上下文、体裁、读者等因素。

3. 功能对等

直译与意译两种途径源于语言差异与文化差异的处理。由此而产生了"强调尽量保持原文形式"与"强调充分体现译文的语言习惯"两种不同的观点和做法。随着翻译研究的深入，关注读者反应为翻译理论与实践提供了一个新的视角，不再是片面强调形式或内容的对等，而是将读者对译文的反应与感受作为衡量译文好坏的重要因素。美国翻译家奈达提出的"功能对等"（functional equivalent，即 dynamic equivalent）就是这一思想的代表。这一原则一改传统，不再将着眼点放在两种语言的对比上，转而强调译文读者的感受，不断地寻找与原文效果"最近"的译法，即在译文所传达的功效上与原文要对等。功能对等把传神性与传意性放在了第一位，在原语与译文差异较大的情况下敢于抛开原文形式，以求在读者感受方面与原文的效果接近。

在译文的表达阶段，除了根据译文的具体情况，恰当地运用翻译方法之外，还必须注意表达的原则性与灵活多样性的辩证关系，即处理好"忠实、通顺"这个翻译标准的两个方面。表达的原则性指的是翻译的忠实性原则，即坚持忠实地传译出原文的思想内容与文体风格。而灵活多样性则指的是在不违背原文的思想与风格的前提上，照顾到译文通顺的一面，采取灵活多样的表达方法，不拘泥字面意思，使译文符合译入语的表达习惯，更好地体现原文思想与风格。坚持原则性是第一位的，如果在这一点上动摇，译文无疑是不合格的，通顺流畅且准确达意的译文是两个方面相得益彰的结果。

在灵活处理译文时，要考虑到译语的表达习惯，例如语篇的连贯与衔接、句序与语序、时态与语态等。只有照顾到这些语言因素，译文才能灵活生动，译文才具有更强的可读性，才有利于思想、风格的传达。在表达时，译者应当有克己意识，也就是克制译者本身的个性，避免望文生义、囫囵吞枣；克制自己的创造欲望，避免天马行空、脱离原文；克制自己的语言定势，避免缩手缩脚、拘泥于字面。翻译是创造性的工作，应根据具体的情况具体对待，在实践中总结和积累翻译表达方面的技巧，才能不断求取进步。

1.6.3 仔细校正

仔细校正是翻译工作的最后一道工序，也是把好翻译质量关的一道重要工序。译者须做好以下几方面的工作：

（1）通读全文，检查译文从整体上是否确切地表达了原文的思想内容，是否有漏译现象；

（2）再一次通读全文，将译文的语言进行润色，使译文在文体方面保持一致，并且检查译文中是否有误译现象，处理表达阶段的遗留问题，如晦涩难懂的译文应根据上下文重新进行梳理，以及调整语言的逻辑关系等，使其符合译入语表达习惯；

（3）检查译文在专有名词、数字、地名方面是否有误译情况；

（4）检查译文的标点用法是否统一，以及译文的文稿编辑方面的特殊要求等；

在做翻译的过程中再仔细也不为过，校正阶段译者也不能掉以轻心，大意了事，只有秉承着仔细再仔细的心态才能产出高质量的译品。

1.7 译者的素质要求

傅雷认为："事先熟读原著，不厌求详，尤为要著。任何作品，不精读四五遍决不动笔，是为译事基本法门。第一要求将原作（连同思想、感情、气氛、情调等等）化为我有，方能谈到迻译。平日除钻研外文外，中文亦不可忽视，旧小说不可不多读，充实词汇，熟悉吾国固有句法及行文习惯。总之译事虽近舌人，要以艺术修养为根本：无敏感之心灵，无热烈之同情，无适当之鉴赏能力，无相当之社会经验，无充分之常识（即所谓杂学），势难彻底理解原作，即或理解，亦未必能深切领悟。"

翻译工作和其他的工作一样，想要做好，必须经历一个勤奋学习长期积累的过程。翻译人员的必备素质概括起来有以下六点：

（1）纯熟的双语能力。纯熟的双语运用能力是做好翻译工作的首要条件。翻译是在两种语言之间进行的转换，涉及对两种语言的语言结构、修辞手段、思维方式等进行分析和比较。双语语言能力的提高能够增强分析语言的能力，从而准确地理解原语。对原语理解准确性的高低也直接影响到表达的准确性和译文的质量。纯熟的双语能力不仅包括理解原语所需的语言能力，同时也包括驾驭译语表达时娴熟的写作能力，这也是众多著名的译者本身就是名作家的原因。

（2）广博的知识。翻译家必须是"杂家"，应具备百科全书式的知识。翻译涉及各个领域，因而须对各国的历史、地理、政治、经济、外交、军事、文化习俗有一定的了解。专业知识的积累与丰富能够为翻译提供背景知识，以提高翻译的准确性。

（3）敏捷的思维能力。思维能力包括逻辑思维（抽象思维）与形象思维能力。敏捷的思维能力是分析和处理双语语言材料的必要条件。能否理清原语的层次关系，能否准确理解作品的思想，能否对于作品的翻译要素进行分析，能否恰当地运用翻译方法等等，这些无不要求译者在翻译时保持敏锐的思维和洞察力。比如翻译政论文章、学术论文以及科技文献要求译者有较强的逻辑思维能力，而翻译诗歌等艺术作品则需要译者有较好的形象思维能力。

（4）高度的责任感。高度的责任感始终贯穿于翻译的整个过程以及翻译工作者的整个生涯，翻译的过程包括理解、表达和校核。周密的准备是保证译文质量的第一步；仔细通读全文，了解原文的相关知识，查阅译者的背景资料，都需要译者认真周密的准备。表达是翻译中的难点，正如翻译名家严复所说的那样："一名之立，旬月踟蹰。"译者的责任心是极为重要的，不应因翻译任务急迫而草率行事。翻译的最后一步是校核。译者应该对文字内容、语言表达、文化风格乃至标点符号进行检查。

（5）敏锐的政治意识。除了具有高度的责任感外，译者还应当具有敏锐的政治意识，这一点应当引起翻译初学者的注意。有时，翻译可能会涉及国家主权、政治立场等问题，正所谓"外事无小事"，译者应当仔细考究，尽可能寻找到出处。

（6）热练掌握翻译理论和翻译技巧。除上述所列的要求之外，倘若译者可以掌握必要翻译理论与技巧，就更加如虎添翼了。理论来自实践，反过来又回到实践中去指导实践。翻译理论也是这样，在翻译实践中不断地被完善更新，同时又为具体的实践提供方向与指南。对于翻译初学者而言，往往会觉得理论技巧与翻译实践有一定的差距，难以结合，但如果能持之以恒，加强实践能力，再结合理论技巧的学习与积累，必能终有所获。

简明建筑英语
翻译教程
Chapter 2

第二部分　翻译中的英汉语言对比

翻译是一种语言活动，是"把一种语言表达的内容忠实地用另一种语言表达出来"的活动，世界上各民族的语言之间有一些共性（language universals）存在，这使得翻译成为可能。语言之间既有共性，更有差异，而正是这些差异又使得翻译存在很多困难。为了更好地理解原文内容并在目的语（target language）中用合适的语言表达原文所传递的意义，本章将从语言层面对英语和汉语之间的异同进行比较。

2.1　词汇

词汇是最基本的语言材料，它是人类语言和生活经验的衔接点。人们若想表达客观物质世界和主观精神世界，首先必须选择恰当的词汇。词汇也是翻译的基本单位，英语和汉语都有丰富的词汇。在英汉两种语言中存在着大量的对应词语；但是由于自然环境、社会历史、文化传统、思维方式的不同，英汉两种语言的词汇在词义、色彩、搭配等方面存在着许多差异。

2.1.1　词义（meaning）

词汇是一种会表示深层概念的符号，与人类对世界的认知方式密切相关。人类对世界的认知具有一定的共性，所以英汉词语存在着一定的对应现象。总的来说，英汉词语在词文方面存在以下几种对应现象：

1. 完全对应（equivalence）

英汉词汇中有一些词语意义是完全对应的，主要是一些专有名词，如：

gene——基因

Chemistry——化学

Paris——巴黎

socialism——社会主义

Beijing——北京

carbon dioxide——二氧化碳

这些词语在翻译时很容易处理，只需一一将其对应的意思翻译出来即可。

2. 部分对应（partial equivalence）

由于人们对客观事物的命名和概念划分方式不尽相同，无论是在英语还是在汉语中，都存在一词多义的现象。所谓英汉词语的对应，绝大多数是部分义项的对应。例如，英语中的 run 与汉语中的"跑"并不完全等同，run 还可以表示"经营、运行"等意义。具体地讲，英汉词义的部分对应可分为如下几种情况：

（1）英大于汉：指一个英语词语可对应多个汉语词语。如：

operate——操作；动手术

sophisticated——世故的；高级的

aunt——姑；姨

cousin——表兄；表弟；表姐；表妹；堂兄；堂弟；堂姐；堂妹

在英译汉的过程中，应根据具体语境来判断英语词语的确切含义。如果要把 a thick slice of bread, thick soup, a thick line 这几个短语分别翻译成汉语，那么译文相应应该是：一片厚面包、浓汤、一根粗线；而 a thin mist 指的是薄雾，thin soup 为稀汤，a thin wire 意为细金属丝。

再看下面这句话：

She has married all her daughters.

其汉译文应为：她把女儿都嫁出去了。

（2）汉大于英：指一个汉语词语可对应多个英语词语。如：

青色——cyan, blue, green

高——high, tall

借——lend, borrow

家——home, family, house

车——car, truck/lorry, motorcycle, jeep, etc.

肉——flesh, meat, pork, beef, mutton, etc.

裁缝——dressmaker, tailor

在汉译英的过程中，应根据具体语境来判断汉语词语的确切含义，并用恰当的英文表达出来。例如：

他鼻梁很高：He has a very tall nose.

一根高高的桅杆：a high mast

再如，"车来了！"这句话若要译为英语，就要判断是小汽车、卡车、吉普车、出租车，还是公交车才能准确地译成 Here comes the car/truck/jeep/taxi/bus！同样，汉语中最普通的问候"吃饭了没有？"翻译成英语时，则要视交际的具体时间意为 Have you had breakfast/lunch/supper?

3. 词语空缺（lexical gap）

由于历史文化、自然环境、社会经济性、文化、科技等各方面的差异，有些汉语词汇英语里没有，而有些英语词汇在汉语中难以找到与之对应的词汇，这就造成了词汇空缺的现象，具体表现为：

（1）汉有英无。

一些中国文化特有的东西，英语里找不到对应的词汇，这就是一种汉有英无的词汇空缺现象，中国的农历节气，阴阳五行、中医、武术、道教、八卦，以及一些传统食品等，在英语中没有对应的表达方式，这就给翻译带来了一定的挑战。如：

阴阳——yin and yang

易经——the book of changes

武术——martial arts, kungfu

饺子——jiaozi, dumpling

烧饼——shaobing, baked pancake

再者就是汉语体系中含有量词，而英语体系里却只有数字，没有量词的词类范畴。如：

一轮明月——a bright moon

一支铅笔——a pencil

一架飞机——a plane

一所学校——a school

(2) 英有汉无。

一些英语国家特有的语言文化现象在汉语中也找不到对应的表达方式，这就造成了英有汉无的词汇空缺现象。例如20世纪初，西方的一些先进理念刚刚传入中国，出现了一些新词，如 science, democracy, 都是汉语中所没有的，最初人们只好将其音译为"赛因斯"和"德莫克拉西"，后来随着社会的发展，它们才逐渐被翻译成"科学"和"民主"。再如：

humor——幽默

logic——逻辑

jeep——吉普车

romantic——浪漫，罗曼蒂克

win – win——双赢

e – mail——电子邮件，伊妹儿

英有汉无的现象还反映在各自的语言体系里。英语中有一类词叫冠词（a, an, the），用来表达特指（definiteness）或泛指（indefiniteness）的概念，而汉语中却没有这类词汇。如：

The sun is red. 太阳是红的。

A book is on the desk. 桌子上有本书。

The book is on the desk. 这本书在桌子上。

"He is a crazy fellow!" said a Colonel Warren. "他是个疯子"一个名叫沃伦上校的说。

2.1.2 色彩（color）

一般来说，词语的色彩包括其中语体色彩、感情色彩、形象色彩、联想色彩等。它是人们在长期的语言使用过程中形成的、附加在词语内涵意义（denotation）上的一种外延意义（connotation）。词语的色彩与社会文化密切相关。英汉两种语言的词汇不仅词义不完全对应，即使意义基本对应的词语，其色彩也存在差异。这里我们主要讨论英汉词语在语体色彩与感情色彩上的差异。

1. 语体色彩（stylistic color）

语体色彩是指词语在某一场合长期使用后而产生的附加意义。词语的语体色彩大致可分为口语体和书面语体色彩。口语包括大量的俚语，它是人们日常生活中使用的语言。书面语多用于科技文章、政论文章、公文等语体中。英汉语中对应的词汇往往都具有各自不同的语体色彩，例如：

mother（书面语），mom, mama, mum, mummy（口语）——母亲、母（书面语）、妈咪、妈、娘（口语）

kill——杀、杀害、杀戮（书面语），宰、干掉、弄死、让……上西天（口语）

在英汉互译中要注意判断原文的语体色彩并在译文中采用相应的语体色彩词。例如：

① It was artful of you, Colonel; but I bare no malice. I should have done the same myself. I've been the victim of one woman after another all of life; and don't grudge you two getting better of Eliza.

上校，你们真鬼，可是咱不怪你：要是咱，咱也这么办。咱一辈子总是吃娘们的亏，咱也不怪你们两位占了伊丽莎的便宜。

显然这段话口语语体特点非常明显，故译者采用了口头语"咱"和俚语"娘们"来传达原文中"I"和"woman"的含义。

② 直起身又看一看豆,自己摇头说,"不多不多,多乎哉?不多也。"

Straightening up to look at the peas again, he would shake his head and reiterate, "No many, I do assure you. Not many, nay, not many at all."(杨完益、戴乃选译)

原文用"多乎哉"这样的古语词来表现孔乙己的迂腐,译文用古英语 nay 实现了译文与原文语体色彩的一致。

2. 感情色彩（emotional color）

感情色彩是词语对客观事物表示感情和态度的附加意义,大体上可以分为褒、贬和中性三种色彩。英汉两种语言中某些基本意义对应的词语往往所传达的感情色彩并不完全一致,如 politics 在英语中含有贬义,而汉语中的"政治"却没有这种色彩,"劳动"在汉语中是一个褒义词,含有高尚的意味,而在英语中 labor 则含有艰苦、令人不愉快的含义。在翻译时一定要注意理解词语的感情色彩并做到正确传达。

从总体上看,英语词语的感情色彩比较模糊,需要根据上下文理解,而汉语词语一般都具有较明确具体的感情色彩。例如:

① But she was so affection and sweet natured, and had such a pleasant manner of being sly andshy at once, that she captivated me more than ever.

她这个小姑娘,感情那样笃厚真挚,天性中温蔼柔和,态度那样羞涩之中含有敏慧,敏慧之中含有羞涩,因此弄得我对她比以前更加倾倒。

sly 这个词既可以表示"狡猾的"（贬义）,也可以表示"淘气的"（中性）、还可以是"灵巧敏捷的"（褒义）。这里把它翻译为褒义的"敏慧"比较切合上下文语境。

英语中还有一些类似的词语,既可以表示褒义也可以表示贬义,如 ambitious——雄心勃勃的/野心勃勃的,aggressive——攻击性的/有进取心的，sophisticated——世故的/成熟的，等等。如：

② Many people think that he is one of the most ambitious politicians of our times.

很多人认为他是现今最有野心的政客之一。

③ Although he is very young, he is very ambitious in his research work.

他虽然很年轻，但是在研究工作中很有雄心壮志。

④ They incited him to go into further investigation.

他们鼓励他做进一步的调查。

⑤ The plotters incited the soldiers to rise against their officers.

阴谋家煽动士兵们造军官的反。

2.1.3 搭配（collocation）

词语的搭配是指词语的共现与组词能力。英汉两种语言由于各自的词汇都存在一词多义的现象，相对应的词汇在搭配能力、搭配范围等方面存在诸多差异。一些词语看似对位，实则不然。例如：

① "开"与"open"。

开门 open the door

开车 drive a car（c. f. open a car）

开会 hold a meeting（c. f. open a meeting）

开井 sink a well（c. f. open a well）

开灯 turn on the light（c. f. open the light）

开枪 fire（e. f. open the gun）

开单子 make out a list（c. f. open a list）

从以上各例可以看出，"开"与 open 的搭配范围并不相同如汉语中可以说"开灯"，而英语中的 open 是不能和 light 搭配的，只能说

turn on the light;"开车"的"开"在英语中只能是 drive,假如"开车"翻译成 open the car,意义就完全改变了。再如:

② "run"与"跑"。

  run a factory 办工厂

  run the block 冲破封锁线

  run the risk of 冒……的风险

  water is running short 水快用完了

③ "wear"与"穿"。

  wear a coat 穿着衣服

  wear a pair of spectacles 戴着眼镜

  wear a beard 留着胡须

  wear perfume 搽了香水

  wear a smile 挂着微笑

以上各例可以看出,英语的 run 搭配范围显然与汉语的"跑"有很大不同;从中可以看出,搭配对于翻译中词义的确定具有重要意义。因此,初学者在翻译时不仅要注意词语的意义,还要注意各词之间的搭配关系及其搭配意义(collocative meaning)。

总之,英汉词汇各自的语义范围不同,在需翻译时要注意词汇意义的准确理解与表达。

2.2 语法(句法结构)

2.2.1 基本语法特征

英语语法和汉语语法最本质的差异可以总结为英语语法是显性的(overt),即可以通过形式看出来词性、语法功能(主语、谓语、宾语等)和语法关系(指代关系、主从关系等),而汉语语法是隐性的

(covert)，即汉语的词句从形式上看不出他们的词性、语法功能和语法关系，具体表现如下：

1. 词性

英语的词性可以表现在形式上，相当数量的英语词具有词性词尾，很多时候我们可以从词尾形式判定一个词的词性，从而判断它在句子中的语法功能，获得句子意义，如以 -ment、-tion、-ism、-ness 等结尾的是名词，以 -en、-ize、-fy 等结尾的是动词，以 -al、-ful、-tive 等结尾的是形容词，等等。

除词尾之外，冠词在英语句子中非常重要，它是标定名词词性的重要标志。标定了名词，还有与之呼应的动词，一个句子的结构就基本定型。

汉语的文字体系与英语迥然不同。汉语的词由字组成，而字多数不具备辨别词性的功能（只有少数汉字可称为词头或词尾，如老-、-化、-性）。很多词在不同的语言环境下具有不同的词性，如：

① 他<u>把</u>那个<u>把</u>门的一<u>把</u>拽住。

这句话里，同样的一个"把"字就有三种不同的词性。在这样的句子中只能靠语义分析、语序和虚词等手段才能确定一个词的词性。再如：

② 君君，臣臣，父父，子子。

这句话的理解依赖于上下文，且仁者见仁智者见智，可以理解为"君要像个君，臣要像个臣，父要像个父，子要像个子"，也可理解为"把君当作君，把臣当作臣，把父当作父，把子当作子"，等等。

在英汉互译，特别是汉译英的过程中，尤其要对词性做出准确的分析，以保证正确传达原文所包含的信息。

2. 时态、语态和语气

英语中动词有时（sense）、体（aspect）、态（voice）、语气（mood）等的变化，都明显地表现在形式上，而且形式比较确定，易于辨认。例如：

① If you go, I'll go too. （时态）

② The writer was assisted in the preparation of his articie. （语态）

③ If you had phoned me. I would not have come to meet you. （语气）

以上例句从动词形式 will go，was assisted 和 would not have come 分别表示将要发生的动作、被动意义及虚拟语气。而汉语中却没有类似的标志，动词在句中都使用原形，英语中对句式起决定性作用的词尾变化，汉语都可以"尽在不言中"。例如：

① 你去我也去。(If you go, I'll go too.)

② 非杀死一头骆驼不可。(A camel must be killed.)

③ 你死了，我就去做和尚。(If you should die, I would go and be a monk.)

上面几个汉语例子实际也分别表达了将要发生的动作、被动意义和虚拟语气，但是从语言形式上却看不出任何标志，时态、语态和语气等在汉语中是隐含的。因此在汉译英中要特别注意对原文的语义分析。

3. 主谓一致

在英语中，一个完整的句子通常要有主语和谓语动词，并且主语和谓语在数和人称方面必须保持一致（SV concord），这是句子构成的基本条件。例如：

① A boy comes. /Two boys come.

② All were damaged, some beyond repair.

③ A lot of money was spent on travel.

而汉语句子的构成机制中就没有主谓一致的规定。在汉语中，句子强调意义而非句子的结构形式，因此主语的地位远不如英语中那么重要，甚至有时可以不出现；而在有的句子中，主语并不明显，理解为"话题"更为合适，因此主谓关系不像英语在形式上要求那么严格，语义的理解更多是依靠上下文和各句子成分的意义。

例如：

① 一点困难怕什么？

② 彩电我们都搬回家了。

以上两句中"多少一点困难"和"彩电"很难按照英语的句子的构成机制理解为"主语"，而是提出一个话题，后面的内容则是对这个话题的进一步说明。

2.2.2 句子结构特征

有学者曾把英语和汉语的句子构造分别比喻为树式结构与竹式结构，这一比喻生动形象，说明了英汉两种语言组织句子的最基本的规律。英语句子有主干（主句）、分枝（从句）与细枝（短语），正如一棵大树；汉语句子往往借助动词，按时间顺序或逻辑顺序层层铺开，像一杆竹子，一节一节连下去。

1. 树式结构

英语的树式结构主要包含三层意思：

第一，句子有一个基本的主干。通过观察囊括一切英语句子的七大句型（SV，SVO，SVC，SVA，SVOO，SVOC，SVOA）我们发现主语 S 和谓语动词 V 是必不可少的成分，不论多么复杂的句子，只要找到主语和谓语动词，就能迅速确定句子的基本框架。

第二，句子的所有枝丫都是从主干上派生出来的。英语句子结构的六大成分中，主语、谓语是不可或缺的主要成分，宾语、补语、定语、状语则是连带或附加成分，就像一棵大树先有一根主干，然后分出较粗的树枝，再分出更细的树枝，但大树的基本格局仍是主干，枝叶对其没有太大影响。

第三，句子的扩展仍是在主干的基础上进行的，不影响基本主干，英语的短语和从句不管多么复杂，都是以不改变句子主干的方式来实现的。下面这首著名的英语童谣就是一个典型的例子：

This is the farmer that sowed the corn,

　　That kept the cock that crowed in the morn,

　　That waked the priest all shaven and shorn,

　　That married the man all tattered and torn,

　　That kissed the maiden all forlorn,

That milked the cow with the crumpled horn,

　　That tossed the dog,

　　That worried the cat,

　　That killed the rat,

　　That ate the malt,

That lay in the house that Jack built.

2. 竹式结构

与英语不同，汉语的句子是竹式结构。汉语句子的竹式结构也包含三层意思：

第一，汉语不存在一个主干结构，没有主干和枝丫之分。从现代汉语来看，有主谓结构的句子和不存在主谓结构的句子大概各占一半，且与英语不同的是，汉语句子没有主谓一致的要求。主干的作用非常重要。

第二，汉语句子的构造方式就像竹子一样是一节一节拔起来的。"竹节"可长可短，可以是词或词组，也可以是短句。

第三，不存在主干的汉语句子，在扩展的时候会引起结构的不断变化。例如：

梅花

一朵梅花

鬓边斜插一朵梅花

红颜小姐鬓边斜插一朵梅花

从上面的例子可以看出，汉语的句子扩展一般是向左的，呈句首开放型；而英语的句子扩展是向右的，呈句尾开放型。

2.3 语序

语序指各级语言单位在组合中的排列次序。汉语属于分析型语言，曲折变化不发达，因此语序是一种极为重要的语法手段。英语属于综合型向分析型转换的语言，有一定的曲折变化，虽然语序在英语中的作用也很重要，但没有汉语那么突出。

从基本语法特征的角度看，汉语语序的强制性比英语强。也就是说，汉语中语序不可轻易改变，往往改变语序就会引起句子成分语法功能的改变。而英语中，这种改变不会引起句子成分的变化。例如：

① a. 他们完成了任务。

　b. 任务他们完成了。

在前一句中，"任务"是"完成"的宾语，句子的重心在于强调"他们"的行为；而后一句中，"任务"成了受事主语，句子的重心变成了对"任务"的陈述。这里，同一词语的句子成分发生了变化，同时，句子的信息重心也随之产生了变化。再看下面的英语例子：

② a. I don't like beer.

　b. Beer I don't like.

这里两句话中 beer 都是 like 的宾语，前一句是正常语序，而后一句是倒装语序。可以看出，在英语中，词语在句中位置的改变并不改变其作为句子成分的语法功能。

汉语语序的强制性还表现在汉语语序改变后，往往会引起意义上或语感上的不合格；而英语语序的改变通常只引起句式的变化，不会导致语感不合格句的产生。例如：

a. 他打点行李动身了。

b. 行李被他打点动身了。

后一句显然是语感不合适，因为不符合汉语的习惯。再看下面的

英语例句,

③ a. I broke the cup into pieces.

b. The cup was broken into pieces.

二者只存在主动句式和被动句式的区别,不会引起意义或者语感上的不合格。

可见,汉语的语序比英语更具有强制性,有时前后稍一改动就会造成意义的重大差别。例如,据说曾经有一位国民党军官犯了错误,上司命令发电报给相关人员处理,他口授了这样的内容:情有可原,罪无可缓。而发报员悄悄改动了两句话的顺序:罪无可缓,情有可原。这小小的改动,却救了这位军官一命。可见语序在汉语中多么重要。

总的来说,汉语的语序比较固定,而英语相对比较灵活。下面我们具体讨论单句和复句层面英汉语序的差异。

1. 单句层面

总体而言,英语和汉语中各句子成分的次序大同小异,一般主语都在谓语动词前,谓语动词在宾语或主语补足语之前,定语在被修饰词之前。但也存在一定差异,主要有以下几个方面:

(1) 主语和谓语动词的位置。

一般情况下,英语或汉语句子中,主语都在动词前,但在疑问句当中,英语和汉语就有了差别。例如:

① Did you have a good time?

你玩得痛快吧?

按照英语的习惯,疑问句里动词在主语之前,不过英语里的动词常常是由两部分组成的,主语放在 do 之类的助动词之后,实意动词之前;而汉语中的语序仍是主语在前动词在后,用语气助词表达疑问。一般疑问句如此,特殊疑问句也是如此,例如:

② What did she say?

她说什么?

除了疑问句之外,英语中还有一些句型,动词是放在主语之前的。如 there，here，then 开头的句子，以及表示祈愿的句子等。如：

③ There must be something wrong with the computer.

④ Here comes your friend.

⑤ Long live our friendship!

而汉语仍旧遵循主语在前动词在后的原则，或者使用无主句。上面几句可以译为：

⑥ 电脑肯定出问题了。

⑦ 你朋友来了。

⑧ 友谊万岁！

（2）宾语的位置。

一般来说，英语的宾语放在主语和谓语动词之后（SVO）；而汉语里除这种语序之外，还有一种特有的宾语在主语和谓语动词之间的句子（SOV），例如：

① 我把作业做完就来。

② 他这个人天不怕、地不怕。

还有一种句子在汉语中也极为常见，即把宾语放在主语之前（OSV），如：

③ 这部电影我上星期看过。

④ 那本书我还没看完。

英语中也有这种语序，但不及汉语常见。主要用于感叹句，或者因为对照而把宾语提前。例如：

⑤ What fine things he has got!

⑥ Many things we gladly remember, others we gladly forget.

（3）定语的位置。

英语的定语可以前置也可以后置，而汉语里定语通常是前置的。对于单个形容词做定语的情况，英语和汉语一般都是定语放在被修饰词之前，英语中也有个别置于被修饰词之后的，如 secretary general。如果是两个或两个以上形容词做定语，或形容词本身带有附加词语，

或者是短语做定语，英语很可能把定语放在被修饰词之后。而汉语句子里的定语成分只能全部放在被修饰词之前。例如，

① I refer to literature pure and simple.

我指的是纯文学。

② This is a problem very difficult to solve.

这是一个很难解决的问题。

③ He is a candidate with little chance of success.

他是个当选希望极小的候选人。

英语和汉语中都有多项定语同时修饰一个中心词的情况。英语大体是按照限定性、描绘性、分类性的次序排列。限定性修饰词的顺序是① all，both 之类的限定词；② 冠词 a，an，the，代词和名词所有格如 my，your，Jack's 等，any，no，every 等；③ 序数词。描绘性定语的顺序是主观评价性词语在前，客观描述性词语在后。其次序是表主观评价形容词——表度、量形容词——表新旧、温度形容词——表形状的形容词——表颜色的形容词。分类性定语顺序如下：表来源（通常是表示国别等）的形容词——表原料等的形容词——表用途等的形容词。汉语中定语的次序大致如下：时间地点定语，领属性定语，限定性定语，短语性定语，国别定语，描绘性定语，本质性定语。试比较：

④ long, curly, red hair.

长长的红色卷发。

⑤ Both the two big round carved French wooden tables in the room are bought yesterday.

房间里这两张法国产的雕花圆木桌都是昨天买的。

另外，英汉地名的书写顺序恰恰相反，这也是定语和被修饰词的语序不同的体现。例如英语说 No. 10 Downing Street, London, England，而汉语则是"英国伦敦唐宁街十号"。

（4）状语的位置。

用作状语的通常是副词和短语。在英语中状语的位置各不相同，

大致可以分为如下几种情况：① 放在主语之前，即句首；② 放在动词前；③ 复合动词的中间；④ 放在动词后，宾语前；⑤ 放在句末；⑥ 放在形容词前。在汉语中，状语通常放在动词或形容词之前，少数以"到""在""给"关联的词语除外，如"说到伤心处""他住在上海"等。

英语中表示程度的状语，单词通常放在被修饰词之前，短语在被修饰词之后。而汉语放在被修饰词之前。例如：

① He is very angry.

② He is angry to a high degree.

而在汉语中这句话恐怕只能表达为：他非常愤怒。

英语中表示状态的状语，如果是单词通常可放在动词之前、复合动词中间或者句末，短语一般只放句末。有时英语表示状态的状语也可置于句首。而汉语的状态状语通常放在动词之前。例如：

① a. He went away quietly.

　b. He quietly went away.

　c. He went away without a word.

　d. Quietly（without a word）he went away.

汉语里通常只能说：他悄悄地走了。如果说成"悄悄地他走了"或者"他走了，悄悄地"都是为了某种修辞效果而特意地改变语序的结果。

英语中表示地点的状语，一般放在句末。偶尔 here 或 there 这类地点状语会位于句首；而汉语中的地点状语通常还是位于动词之前。例如：

② He's having breakfast in a small restaurant.

他在一个小餐馆吃早餐。

③ Here he wrote that great novel.

在这里他写出了那部伟大的小说。

英语中表示时间的状语，通常放在句末，但也可以放在句首；汉语的时间状语通常放在句首或动词之前。例如：

④ a. He left for Beijing yesterday.

　　b. 他昨天动身去北京了。

⑤ a. Yesterday he left for Beijing.

　　b. 昨天他动身去北京了。

英汉互译时，可以根据两种语言各自的特点进行语序的调整。特别是汉译英时要根据英语的特点灵活处理主语、谓语、定语、状语等的位置。

2. 复句层面

复句即复合句，包括并列复合句和主从复合句。由于英语的语法是显性的，复句中分句之间的并列或主从关系主要靠关联词语来体现，如定语从句、状语从句、主语从句、宾语从句等，各分句的顺序是由语言形式控制的，语序比较灵活。汉语复句中分句间的关系主要靠逻辑关系联接，逻辑顺序受比较严格的规律制约。具体来说，在语言的组织上要有以下几种顺序：时间顺序、大小顺序、因果顺序、重轻顺序。这些规律使汉语的语序相当固定。例如：

① I had spent a long day on a hired mule before the mail carrier who had been my guide pointed to a cabin on the far side of a stream, mutely refused the money I offered, and rode on.

我租了一头骡子，邮差权充向导，骑了一整天，然后他遥指着河那边的一幢木屋。我给他钱。他默然拒绝，径自骑骡走了。

这句话英语原文包含多个动作，它是以 before, who 和 and 这些词来连接各个或主从或并列的分句，形式严密，但译成汉语就必须依照汉语的习惯，按时间的先后顺序重新组织，若按原句顺序翻译过来的话一定诘屈聱牙，读来很别扭。

② Strolling unescorted at midday past a major concentration of the huts just a block from the city's Central Avenue, I nonetheless saw many signs of occupation.

中午，我在没有导游陪伴的时候独自漫步街头，在中央大道附近发现了一个很大的棚户区，很多茅棚里还住了人。

此句有四个表示处所的词语，在汉译时按照汉语中范围从大到小的顺序，把"中央大道"这一处所前移，比较符合汉语的语序习惯，流畅自然。

再看下面汉译英的例子：

③ 我看见他戴着黑布小帽，穿着黑布大马褂，深青布棉袍，蹒跚地走到铁道边，慢慢探下身去，尚不大难。可是他穿过铁道，要爬上那边月台，就不容易了。

I watched him hobble towards the railway track in his black skullcap, black cloth mandarin jacket and dark blue cotton - padded cloth long gown. He had little trouble climbing down the railway track. but it was a lot more difficult for him to climb up that platform after crossing the railway track.

上面划线部分例句中，汉语的语序是按照时间顺序先后排列，译成英语要变为主从结构，用 after 连接，可见英语有严密的联接手段，关联词语在确定句式方面更加重要，而汉语组句一般遵循意义的连接。

④ 我告诉他，家有老母长年患病，我离国已六、七年，想回去看老母，至多两年就出来。

I told him that I had been abroad for six or seven years and that I had to go home to see my old mother, who had been ill for a long time. However, I assured him that I would come back (at most) in two years.

例④中，汉语原文各分句是按照因果顺序来列的：先有"母亲患病"以及"我离国多年"的原因，然后才说到"想看老母"；而英译文中把"想回去看老母"与"离国已六、七年"并列作为宾语从句，把"长年患病"处理为定语从句来解释说明"老母"的状况，结构紧凑，改变了原文的语序，但非常符合英语的组句习惯，读起来连贯自然，逻辑性强。

汉语语序按照时间、大小、因果、重轻排列的规律不仅适用于汉语复句层面的语序规律，也适用于总结单句以及短语、词汇的语序特征。例如：

① 自打离开上海我再也没见过他。

I had never seen him since I left Shanghai. （时间）

② 大中小城市

small, medium-sized and large cities （大小）

③ 花快干枯了，他浇上了水。

He watered the flower because they were dry. （因果）

④ 中国古代炼金术

the ancient Chinese alchemistry （重轻）

2.4 修辞

本节讨论的是广义的修辞，即遣词造句，不包括积极修辞，即辞格的运用。总的来说，英语和汉语在修辞层次上的差异主要表现为以下几点：

2.4.1 形合与意合 (Hypotactic vs. Paratactic)

形合指借助语言形式手段（如关联词等）实现句子的连接，表达语法意义和逻辑关系。意合指不借助语言形式手段而借助词语或分句的意义或逻辑联系实现句子的连接。英语重形合，注重形式衔接（cohesion），要求结构完整，以形寓意，严密规范。汉语重意合，注重意念连贯（coherence），不求结构完整，以意统形，流泻铺排。例如：

① If winter comes, can spring be far away?

冬天来了，春天还会远吗？

② 知己知彼，百战不殆。

You can fight a hundred battles without defeat if you know the enemy as well as yourself.

1. 英语的形合法

具体来说，英语句子的形式连接手段（cohesive ties）主要有以下几种：

（1）关系词与连接词。

关系词包括 who, whom, whose, that, which, what, when, where, why, how 等用来连接主句和定语从句、主语从句、宾语从句或表语从句等的代词、副词。连接词包括并列连词和从属连词，如 and, or, but, yet, so, however, as well as, when, while, as, since, unless 等。这些关系词和连接词几乎是英语句子必不可少的成分，而汉语则很少用这类词甚至不用。如：

① A gentle breeze swept the Canadian plains as I stepped outside the small two-story house.

我步出这幢两层小屋，加拿大平原上轻风微拂。

② When I try to understand what it is that prevents so many Americans from being as happy as one might expect, it seems to me that there are two causes, of which one goes much deeper than the other.

为什么如此众多的美国人不能像想象中那样幸福呢？我认为原因有二，而二者之间又有深浅之分。

（2）介词。

介词是英语里最活跃的词类之一，是连接词语或从句的重要手段。介词包括简单介词如 in, to, of, with 等，合成介词如 upon, unto, within, throughout 等，以及成语介词如 along with, with regard to, apart from 等。英语句子离不开介词，而汉语常常不用或省略介词。试比较：

While the present century was in its teens, and on one sunshiny morning in June, there drove up to...large family coach, with two fat horses in blazing harness, driven by a fat coachman in a three-cornered hat and wig, at the rate of four miles an hour.

当时我们这个世纪刚开始了十几年。在六月里的一天早上，天气晴朗，契思克林荫道上平克顿女子学校的大铁门前面来了一辆宽敞的私人马车。拉车的两匹肥马套着雪亮的马具，肥胖的车夫戴了假发和三角帽子，赶车子的速度不过一小时四英里。

（3）词语的形态变化。

词语的形态变化包括词缀变化，动词的时、体、语态、语气，形容词和副词的比较级，名词的性、数、格，代词的人称等。例如：

① This is a legacy he had left to me, of inestimable value, which we could not buy, with which we cannot part.

这是他留给我的无价之宝，当时用金钱买不到，现在也不能丢掉。

② Production cost has been greatly reduced.

生产成本大大降低了。

（4）定式结构。

英语中还有许多定式复杂句的结构框架也是一种形合手段，如it is...to...句型，there...句型。例如：

① It is necessary to know grammar, and it is better to write grammatically than not, but it is well to remember that grammar is common speech formulated.

懂得语法，很有必要。写得符合语法当然比不合语法好——但要记住：语法只是日常言语的公式化。

② There were 10000 students in our school last year.

去年我校有一万名学生。

2. 汉语的意合法

汉语的意合法主要采用以下手段：

(1) 语序。

汉语中很多并列和主从复句都不用关联词,但并列和主从关系却很清晰。以因果关系主从句为例,汉语中多数因果句都是先因后果,例如:

① 人不犯我,我不犯人。

We will not attack unless we are attacked.

② 小王今天身体不舒服,别指望他干活了。

Xiao Wang is out of sorts today, hence he is not expected to work.

(2) 词句整齐的句式。

词句整齐的句式常用的手段有反复、排比、对偶、对照等。例如:

① 种瓜得瓜,种豆得豆。

As a man sows, so he shall reap.

② 往事不可止,来者犹可追。

Although we cannot undo the past, we can map out our future.

③ 东边闪电出日头,西边闪电必有雨,南边闪电天气热,北边闪电有雷雨。

If it lightens in the east, it will be sunny; if it lightens in the west, it will be rainy; if it lightens in the south, it will be sultry; if it lightens in the north, it will be stormy.

(3) 紧缩句及四字格。

紧缩句就是由复句紧缩而成的,即取消各分句间的语音停顿,略去原来分句的一些词语,使之更加简约。四字格是汉语里广为运用的语言形式,往往言简意赅,简洁明了。例如:

① 不到黄河心不死。

Until all is over, ambition never dies.

② 不进则退。

Move forward, or you'll fall behind.

相较之下,英语更加注重形合,句子结构形式比较严谨;而汉语注重意合，语言比较简洁。因此，在英译汉时往往要先分析句子的结构形式，然后确定句子的意义、功能。而汉译英时往往要先分析句子的功能意义才能确定句子的结构形式。如：

① You try to persuade him now, I talked to him all last night, therefore I've done my part.

你去说服他吧，我昨晚跟他谈了一夜，尽力了。

② There have been many mythical stories about Yu's birth. One is that three years after Gun was killed, his body still showed no sign of putrefaction, and when someone cut it open, out bounded Yu the boy. Another has it that Yu's mother give birth to him after eating a kind of wild fruit. Anyway, in ancient times, everyone seemed to believe that Yu was the son of a god, an ingenious, capable and peerless hero.

关于禹的出生有许多神奇的传说。有的说，鲧死了三年，尸体还没有腐烂，有人用刀子把尸体剖开，禹就跳了出来；有的说，禹的母亲吃了一种野果，就生下了禹。大家都说禹是神的儿子，是一个聪明、能干、了不起的英雄。

2.4.2 静态与动态（Stative vs. Dynamic）

英语倾向于用名词，名词不管是抽象还是具体，都指代具有稳定性质的实体，以静态为特征，因而以名词（和介词）占优势的英语，叙述呈静态；汉语倾向于用动词，动词表示行为、活动或变化状态，以动态为特征，因此汉语中的叙述呈动态。

1. 英语的静态特征

（1）名词化（nominalization）。

名词化是英语中常见的现象，是指主要用名词来表达原本属于动词或形容词所表达的概念，如用抽象名词来表达动作、行为、变化、状态、品质、情感等概念。在许多情况下，这种名词优势可以使行文表意更为简洁、灵活。其具体表现有以下几种：

以名词表达动词的概念分为两种，一种是用抽象名词表示行为和动作等概念（如以 - ment，- tion，- al 等结尾的名词），另一种是用含有行为和动作意义的普通名词代替动词（如以 - er，- or 结尾的名词），将动词转化为名词乃是英语中极为普遍和有效的构词方法，名词与动词的相互转化正是英语的独特之处。例如：

① His dismissal was a good riddance.

他被解雇了，真是天大的好事。

② He is a good eater and a good sleeper.

他能吃能睡。

以名词代替形容词，使语义结构简化。这种结构表达方便简炼，词数少而信息量大，在现代英语中随处可见，在科技英语中尤为常见：

job opportunity discrimination　就业机会歧视

pressure cooker　高压锅

space shuttle flight test program　航天飞机试飞计划

名词优势随之带来介词优势。英语在漫长的发展中逐渐简化了其名词和形容词的形态，促进了介词语法功能的加强，名词的"格"的语法功能，现代英语都由介词来实现，例如，介词 of, for, from, to 等。介词优势与名词优势二者相辅相成，使英语的静态倾向更为显著。例如：

③ The abuse of basic human rights in their own country in violation of agreement reached at Helsinki earned them the condemnation of feedom - loving people everywhere.

他们违反在赫尔辛基达成的协议，在国内侵犯基本人权，因此受到了各地热爱自由的人们的遣责。

（2）动词的弱化与虚化。

现代英语中最常用的动词是 to be，而它也是动作意味最弱的动词，若与 it 或 there 连用则构成的句式静态意味更加明显。试比较：

① There was a tropical storm off the east coast of Florida.

② A tropical storm lashed the cast coast of Florida.

其他动词如 have，grow，get，become，do，go 等也是英语中的弱式动词，其动作意味不强，更具静态感。例如：do some washing，do a little sailing，go shopping 等。

此外，由于英语常常把一些动词转化或派生成名词，这些名词往往要放在一些虚化的动词之后如 take a walk，make attempts，have a look，do some damages 等。这类动词短语往往显得平淡无味，缺少动态感。例如：

③ An attempt must be made to save the pandas.

（3）用形容词或副词表达动词的意义。

英语常用与动词同源的形容词来表达动词的意义，因为有不少动词可以用同源的形容词来表达，如 thank——thankful，create——creative，sympathize——sympathetic，cooperate——cooperative 等。例如：

① I am very thankful to you because you are very cooperative.

你配合的很好，我非常感谢。

英语中表示生理或心理感觉的形容词也可以表达相当于动词的意义。如：

② He was unaware of his father's presence.

他当时不知道他父亲在场。

英语的副词也可以用来表达动词的意义，如：

③ He'll be home soon.

他很快就会回家。

④ The Dallas Mavericks had won several matches, but it was out in the semi-finals.

达拉斯小牛队赢得了多场比赛，但却在半决赛中出局了。

2. 汉语的动态特征

(1) 汉语动词可以充当句子里的各种成分，并且形式不发生变化，例如：

学不嫌晚。（动词作主语）

我爱中国。（动词作谓语）

他们爱学习。（动词作宾语）

革命不是请客吃饭。（动词作表语）

她咯咯地笑起来了。（动词作补语）

他不停地来回走动。（动词作状语）

汉语的动词可以充当助动词，位于动词之前或动词之后。前置助动词主要有来、去、能、会、要、肯、愿意、应该等。后置助动词主要有来、去、起来、上去、过去等。例如：

① 我来叫您。

② 走自己的路，让别人说去吧。

汉语的动词还可以充当介词，有些动词和介词可以互用。实际上，汉语的介词多是由动词演化而来。例如：

用人有方（动词）；用左手写字（介词）

她在家（动词）；她在家看书（介词）

(2) 汉语的动词常常可以连用。动词连用形成了连动式、兼语式、"把"字句、"被"字句等典型汉语句式。例如：

① 我到办公室去找经理。（连动式）

② 我把李教授请来做报告。（把字式＋兼语式）

③ 他叫你把那床毯子抱出去掸掉灰。（兼语式＋把字式＋连动式）

汉语动词常常重叠或重复，构成丰富多样的动词组合。这些组合可以明显地加强汉语动态感的表现力。例如：

① 让我看看。（AA 式）

② 说说笑笑，跑跑跳跳，孩子们过得十分愉快。（AABB 式）

③ 这个问题大家来讨论讨论。（ABAB 式）

2.4.3 被动与主动（Passive vs. Active）

被动语态在英语中是一种常见的语法现象，被动意义常常通过形式即被动语态来表达，特别是在科技、法律等正式文体中，被动句的使用更加广泛。而汉语常用主动形式来表达英语的被动意义。这就是英汉语被动与主动的区别。试比较：

My first thirty years were spent in Western Canada.

我在加拿大西部度过了人生前三十年时光。

1. 英语的被动特征

被动语态在英语中应用相当广泛，尤其常用于科技文章、报刊文章、官方文章等。在这些文体中，使用被动句几乎成了一种表达习惯。英语常用被动句，主要原因有以下几种：

（1）施事。

人们在表达思想时常使用主动句。但当动作的实施者即施事者（agent）不需要指明或者根本不可能指明时，英语往往使用被动句：

当说话者不知施事为何人，一般用被动句，如：

① The traffic problem in Beijing can be solved by developing public transportation.

当施事无需提及，或者说话者避免提及时用被动句，如：

② Why should the unpleasant jobs be pushed on me?

当需要强调受事者，或者施事不如受事重要，要用被动句，如：

③ My computer was broken yesterday, so I can't get online.

（2）句法或修辞。

英语重形合，注重句法结构和表达形式。当主动式不便表达时，出于造句的需要或修辞的考虑，往往采用被动式。

为了使句子达到更好的平衡，以符合英语句子中的尾重原则（end weight），即主语简短而谓语可以较长的习惯，一般采取被动表达方式，如：

① The view was adopted by those who are fighting for their emanipation.

为使句子前后连贯，便于衔接，往往采取被动形式，如：

② He went to the room and was met by two policemen.

有时为避免句型单调，使表达方法灵活多变，达到一定修辞效果，可使用被动式，令句子干脆、有力，如：

③ The basic English sentence pattern of subject can be varied in many ways.

（3）文体。

科技文体、新闻文体、公文文体和论述文体较多使用被动句，以迎合其表达的需要。科技文体注重实例和活动的客观叙述，力戒作者的主观臆断，因而常常避免提及施事；新闻报道注重客观公正冷静的叙述，所以施事者不宣言明；公文则注重叙述公正无私，口气客观正式，因此也常用被动式。如：

As oil is found deep in the ground its presence cannot be determined by a study of the surface. Consequently, a geological survey of the underground rock structure must be carried out. It is thought that the rocks in a certain area contain oil, "drilling rig" is assembled. The most obvious part of a drilling rig is called a "derrick". It is used to lift sections of pipe, which are lowered into the hole made by the drill. As the hole is being drilled, a steel pipe is pushed down to prevent the sides from falling in.

2. 汉语的主动特征

汉语通常用主动的形式表达英语被动式所包含的意义，有些语言学家称之为"隐性被动"。汉语中被动意义主要有以下几种表达方式：

（1）使用"受事主语+动词"句型。

按中国人的思维习惯，人的行为必须由人来完成，事物不可能完成人的行为。所以人们在表达时常常把施事者隐含起来,把注意力集中在受事者及行为本身,因此,受事者便充当了主语。汉语中许多表示行为的动词既可以表达主动意义,也可以表达被动意义。所以,"受事主语+动词"的句型在汉语中很常见,其被动意义是交际者的语感共同认知的。例如:

① 这件事要再研究一下。

The matter must be further discussed.

② 昨晚我盖了两条被子。

Last night I was covered up with two quilts.

(2) 使用无主句或主语省略句。

当主语不明或不重要或需要避免提及时,汉语中的无主句也可以用来表达被动意义。而英语的句子注重结构齐整,每句话必须有主语,只好采用被动句或其他句式把施事者省略。例如:

① 禁止砍伐树木。

Tree felling is forbidden.

② 不努力工作就不会成功。

One can never succeed without making great efforts in one's work.

(3) 采用通称泛称做主语。

这种通称或泛称主语常用于施事者难以指明或者是不便于采用被动式的情况,通常包括"有人""人""人们""大家""我们"等。英语中也有不定代词如 one,some,somebody,anybody 等,但英语通常采用 it 作形式主语的被动句来表达相应的意义。如:

① 大家知道,电子是极为微小的负电荷。

It is known that electrons are minute negative charges of electricity.

② 有人向他指出表格填得不对。

It was pointed out to him that the form hadn't been properly filled in.

(4) 用"被字句"等被动句式。

必须指出,汉语的被动式使用受到句法和语义的限制,用途非常有限。英语中主动句大多可转换成被动句，而汉语主动句大多数不能转成被动句。例如：

① The ceremony was abbreviated by rain.

因为下雨，仪式缩短了（不能说"仪式被雨缩短了"）。

汉语的"被"字从"遭受"的意义演变而来，主要表达对主语而言不如意的事，而其他表达被动的词如"给""叫""让""受""遭""为……所……"等也多表达不如意或不企望的事。例如：

② 这件事给弄糟了。

It was poorly done.

③ 这家工厂在地震中遭到严重破坏。

This factory was seriously damaged during the earthquake.

汉语的被动式不仅受意义的限制，还受到句法结构的限制，因为被字后面一般要有宾语来表示施事者，所以很多难以说出或者不便指出施事者的句子不能变成被动句。如：

④ The sports meeting will be held next Wednesday.

运动会将于下星期三召开（不能说"运动会将于下星期三被召开"）。

（5）其他转换形式。

其他转换形式包括把字式、将字式、"是……的""……的是"式、"加以予以/抱以"等。例如：

① 隆隆的大炮声把傀儡军吓坏了。

The puppet soldiers were frighted to death by the rumbling of cannons.

② 利用发电机，可以将机械能再变成电能。

The mechanical energy can be changed back into electrical energy by a generator.

③ 历史是人民创造的。

History is made by the people.

④ 推荐我的是一位教授。

I was recommended by a professor.

⑤ 他出现在大厅门口时，观众报以热烈的掌声。

He turned up at the entrance of the hall and was warmly applauded by the audience.

2.4.4 物称与人称（Impersonal vs. Personal）

英语常用物称表达法，即不用人称来叙述，让事物以客观的口气呈现出来。而汉语常常倾向于描述人及其行为或状态，因而常用人称表达法。这一特点主要表现在使用主语和谓语动词两方面。

1. 英语的物称表现

物称表达法是英语中常见的表达方式，尤其常见于书面语，如公文、新闻、科技论著和文学作品这种表达法往往使叙述显得客观公正，结构趋于严密紧凑，语气较为委婉、间接，主要表现是使用非人称主语和被动语态。

（1）英语常用非人称主语表达，注重"什么事发生在什么人身上"，非人称主语句主要有两大类：

一类是用抽象名词或无生命的事物名称做主语，同时使用本来表示人的动作或行为的动词做谓语。这种句式往往带有拟人化的修辞色彩，反映了英美国家的幽默感。这类主语常常被称作"无灵主语"（inanimate subject），表示抽象概念、心理感觉、事物名称或时间地点等，但其搭配的却往往是"有灵动词"（animate verb）。例如：

① Astonishment, apprehension and even horror oppressed her.

她感到心情抑郁，甚至惊恐不安。

② The year 1949 saw the founding of the People's Republic of China.

1949年，中华人民共和国成立了。

另一类是用非人称词语 it 做主语。It 可用来替代人以外的有生命或无生命物，也可以做先行词（anticipatory "it"）、强调词（emphatic "it"）或者用作虚义词（unspecified "it"）代替主语。例如：

③ It has never occurred to me that he was so dishonest.

我从没想过他这么不诚实。

④ It's only ten minutes' walk to the bus stop.

到公交站只有十分钟的路程。

（2）英语常用被动式，采用物称表达法。尤其是 it 做主语的非人称被动式（impersonal passive），往往能把所叙述的事实或观点客观、间接、婉转地表达出来，常见于公文文体、科技文体及新闻文体，这类句式包括：

It is believed that…据认为……

It is reported that…据报道……

It is estimated that…据估计……

It is suggested that…有人建议……

It should be realized that…必须认识到……

It must be pointed out that…必须指出……

It is stressed that…人们强调……

It is considered that…大家认为……

It can't be denied that…不可否认……

2. 汉语的人称表现

汉语比较注重主体思维，这种思维模式往往从自我出发来理解、演绎、描写客观环境这个外在世界中的事物，或倾向于描述人及其行为或状态，因而较常用人称。这一特点主要表现在：

（1）使用人称主语。

这里的人称主语不仅仅指人称代词做主语，而是指以人为主体的主语。因为汉语比较注重"什么人怎么样了"，所以常用人称主语表达，例如：

① 他为人和善,因而朋友很多。

His bonhomie often brought him many friends

② 我突然想到一个好主意。

A good idea suddenly struck me.

③ 近年来热情的读者纷纷致函各地方报纸,对本市的城市建设提出了各种建议。

In recent years local newspapers have been sprinkled with impassioned letter advancing various suggestions on the city's urban construction.

(2) 多用主动形式,少用被动式。

中国人的思维习惯重"事在人为",动作和行为必须由人这一主体进行或完成,事或物不可能自己去进行或完成任何行为或动作。所以,若要说出施事者就用人称表达法,若无法说出确定的人称就用"有人""人们""大家"等这些泛称,若无法采用泛称则用无主句,若人称不言而喻又常常省路人称。例如:

① 他开车时心不在焉,几乎闯祸。

His absence of mind during the driving nearly caused an accident.

② 我们知道,物质占有空间。

Matter is known to occupy space.

③ 发现了错误,一定要改正。

Mistakes must be corrected when they are found.

简明建筑英语
翻译教程
Chapter 3

第三部分　翻译中非语言层面的对比

3.1 思维

思维方式与语言密切相关。语言大师吕叔湘先生曾指出"思维方式、思维特征和思维风格是语言生成的哲学机制。语言实际上是紧紧附着在思维这个有无限纵深基础上的结构体"(吕权湘 1990)。由此可见,语言是思维的载体和思维的主要表现形式,而思维又是语言生成和发展的深层机制。不同的语言群体具有不同的思维方式、特征和风格,这些差异必然会在语言结构中表现出来,所以思维方式的差异正是造成语言差异的一个重要原因。翻译不仅是语言形式转换的过程,而且是思维方式变换的过程。要研究英汉语言之间的转换,就必须深入研究与语言文化密切相关的思维方式。下面我们将概略地阐述汉英思维方式的差异及其在语言上的表现。

3.1.1 整体思维与分析思维

整体性思维与分析性思维是人类思维的两种基本形式。整体性思维把天地、人和自然、社会、人生放在一起从整体上综合考察其有机联系,注重整体的关联性,注重用辩证的方法去认识多样性的和谐和对立面的统一。分析性思维明确区分主体与客体人与自然、精神与物质、思维与存在、灵魂与肉体、现象与本质,并把两者分离、对立起来,分别对这个二元世界作深入的分析研究,注重从事物的本质来把握现象。由于传统文化的影响,东西方形成了"东方重综合(即整体),西方重分析"思维习惯。

1. 汉语民族的整体思维

中国哲学强调思维上的整体观,阴阳太极图的标志就很好地体现了中国的整体辩证思维。两端互补,相反相成,相灭相生。先人从自然万物的关系中产生了"天人合一""万物一体"的意识，经过上千年的发展，有机整体性已成为中国传统思维方式的大特性。这种整体性思维方式对汉语的表现方式有着极深的影响，主要表现为以下几个方面：

（1）整体观体现的造字构词法。

汉语造字构词注重统一观，具体表现为造字或构词时先确定类属，然后个别区分。例如汉字中的形声字，以部首来统率，与树水有关的都从"木"字旁，如"松、柏、梅、柳、桦、树、林"等；与水有关的皆从"水"旁，如"江、河、湖、海、溪、池、深、浅"等。这就是一种整体思维的体现每个字都不是孤立的，而是整体系统的一部分。后来词汇发展了，由单音节词发展到双音节和多音节词，但构词时应遵循先确定类属、再加以区分的方法。例如汉语首先把事物本身表面或外面包的一层东西统称为"皮"。然后分别给予描写，如"人皮""树皮""兽皮""书皮"等；再如把木本植物通称为"树"。然后把不同的树分别称为"松树"。"柏树""桃树""柳树""橡树""白杨树"等。英语中就没有汉语这种统一观照的命名方式，如前面所说的各种树，英语分别叫做 pine、cypress、peach、willow、oak、white poplar 等。

（2）语序上的整体到部分。

整体思维常从全局出发，从整体到部分，强调整体的平衡和统一感，这种整体到局部整体思维模式在汉语表现手法上体现为：

在时间上，汉语单位从大到小，即年、月、日、星期、时、分。例如人们常说 2007 年 6 月 14 日 7 点 50 分，决不会倒过来说 50 分 7 点 14 日 6 月 2007 年；

在空间上，汉语是从大到小，从宽到窄，从远到近，从上到下，从整体到局部，例如地址的写法是：北京市朝阳区吉庆里小区 9 号楼 A 座 1002 室；

在介绍人物时，常先列出头衔后点名，然后从大到小依次列出职务，例如：

中国共产党中央委员会委员、中国共产党中央政治局委员、中国共产党中央委员会政治局常务委员会委员，中华人民共和国国务院总理周恩来同志于1976年1月8日不幸逝世。

在叙事时，汉语基本上是从大范围到小范围、由重要意义到次要意义。例如：

① 山里有座庙，庙里有个老和尚。

② 救死扶伤。

③ 那些无房户和拥挤户。

（3）句子安排的整体性重复。

整体的统一感也体现在汉语的句子安排中，汉语经常出现回环性重复（也称周遍性重复），这也是中国式整体思维风格的表现。中国古代就有"鱼戏莲叶东，鱼戏莲叶南，鱼戏莲叶西，鱼戏莲叶北"的诗句，这是典型的周遍性整体重复。而英语中就没有这种表达法，例如：

镇子坐落在一个山谷里，东面是山，西面是山，南面是山，北面也是山

这句话只能译为：

The small town lie in a valley surrounded by mountains.

回环重复使汉语句子很有节奏感和形式美。由上例可以看出，汉语常采取全程演绎式的铺叙法，而英语则多用归纳式。

（4）语段结构的浑然一体。

整体思维方式也表现在汉语语段结构中，其特点是流泻式铺排，主调难分、主从难辨，似乎一切信手拈来，自然随意，浑然一体。例如：

接着，他继续设想，鸡又生鸡，用鸡卖钱，钱买母牛，母牛繁殖，卖牛得钱，用钱放债，这么一生事的发财计划，当然也不能算是生产的计划。——《燕山夜话》

2. 英语民族的分析思维

西方哲学以分析思维占优势，注重严密的形式论证，它大大促进

了西方自然科学的发展。分析性思维强调理性分析，注重个体的独立性，因而形成一种强调以经验为基础，着重形式道科论证的思维定势。表现在语言上是不求全面周到，但重结构上的严谨性。

（1）词形变化。

分析型的思维模式使英语具有明显的词形变化，从而产生了形式多样的语法形式和较为灵活的语序结构。英语组词造句主要靠词形变化。例如：

tradition—traditional

friend—friendly

regular—irregular

architect—architecture

英语中构词主要用加缀法，词缀的增加带来词性的改变，而词性非常重要，因为它将决定其在句子中的成分，例如：

Landing on the loose, fist-size stones alongside the track, he had to struggle to keep his balance.

这句话中除保留一个主要动词外，其他动词分别变成分词和不定式来说明动作之间的先后关系和因果关系。译成汉语的话就要按时间的先后顺序了，他落在铁轨旁拳头大小的散石上，使尽力气才保住平衡。

（2）语序上的部分到整体。

分析型思维往往从局部出发，从部分到整体、强调形式结构程式。其表现为：

在时间上，从小单位到大单位，即分、时、星期、日、月、年，如2007年6月7点50分，英语应该表达为：at ten to eight a.m., on the 4th of June, 2007。

在空间上，从小到大排列。例如"北京西城区展览路1号"，英语应该表达为：No. 1 Zhanlanguan Rd., XichengDistrict, Beijing.

在叙事上，从小范围到大范围，从次要意义到重要意义。例如：

heal the wounded and rescue the dying; the families who have a small house or no house at all.

（3）语段结构的环环相扣。

分析思维体现在语段安排上就是英语的语段结构是分析型多层环扣式语段结构，即并列或者主从关系从形式上明显地表现出来，整个语段中的各个成分用各种连接词体现主从关系或者并列关系。例如：

This instrument works on the principle that each individual substance emits a characteristic spectrum of light when its molecules are caused to vibrate by the application of heat, electricity, etc. and after studying the spectrum which he had obtained on this occasion, Hillebrand reported the gas to be nitrogen.

3.1.2 具体思维与抽象思维

具体思维是指头脑对形象思维分析的活动。它是利用具体的形象素材来集中地再现客观存在，以反映基本的质与规律。抽象一词源于拉丁文，原意指分离、排除或抽出。抽象思维使人们在认识过程中借助于概念、判断和对立的思维形式反映客观事物。从总体上看，汉文化思维模式倾向于具体，常常以具体的形象表达抽象的内容；而西方思维模式具有较强的抽象性，多使用抽象表达法。

1. 汉民族的具体思维偏向

中国传统哲学在注重"玄览"的同时，有倡导重具体、重物象的思维风格。传统思维模式极为重视以物象来体现或比喻抽象的事物。这种思维风格表现在汉语中主要是在表达上"实""明""直""显""形""象"，即措辞具体、涵义明确、叙述直接，常常借助于比喻和形象，比较平易朴实（down to-earth style），汉语表达上的具体主要表现在以下几个方面：

（1）用词倾向于具体。

汉语常用实的形式表达虚的概念，这是因为汉语缺乏像英语那样的词缀虚化手段，需要用具体的词语表达抽象的概念。试比较：

① 那时他们最渴望的就是结束这摇摆不定的局面。

What they wanted most a that time was an end of uncertainties.

② 任何国家都不能自称一贯正确。

No country should claim infallibility.

（2）大量使用动词而不用抽象名词。

汉语使用动词的倾向在前面我们已经做过论述。由于汉语中动词优势的特点，动词在使用时非常灵活、方便，英语中常用抽象名词的地方汉语常常用动词，看下面这句话：

① Starvation was a remote threat.

要把它译成地道的汉语，我们可以说：他们一时不必担心饿死。但如果说"饥饿是个遥远的威胁"就非常别扭，不符合汉语的习惯。

在一些公示语中，汉语倾向于使用动词，如：

② 谢绝参观！

③ 非公莫入！

这两句译为英语就要用名词结构：

② Inspection declined.

③ No admittance except on business！

（3）使用形象性词语。

汉语中形象性词语相当丰富，有大量的比喻、成语、谚语、歇后语等，这些词语的使用使汉语在表达法上更有具象的特点，从以下各例中可以看出来：

远见卓识　　far‐sightedness

伶牙俐齿　　eloquence

水乳交融　　perfect harmony

望穿秋水　　await with great anxiety

下列句子也是很好的证明：

① 他等着她来，急得像热锅上的蚂蚁。

He waited for her arrival with a frenzied agitation.

② 他这一阵心头如同十五个吊桶打水，七上八下，老是静不下来。

His mind was in a turmoil these days and he was quite unable to think straight.

（4）使用丰富的量词。

汉语善用具象的思维风格还表现在量词的丰富多样中。不论是具体的还是抽象的事物，汉语都倾向于把它们度量化、单位化，例如：一支笔、一朵花、一匹马、一寸光阴一寸金，等等。

2. 英美国家的抽象思维偏向

抽象思维的特点是能从纷繁复杂的表面现象中概括事物的本质，并从个别上升到一般，能从总体上认识和把握事物。英美国家把抽象思维看作是一种高级思维（superior mind），是文明人的一种象征（mark of civilized man），抽象模糊的意义也迎合了人们的某种表达需要，因此抽象表达法成为一种流行。英语的抽象表达主要在于大量使用抽象名词。

另外，因为英语有丰富的词义虚化手段，所以大大方便了抽象表达法的使用。这些手段主要有：

（1）用虚化词缀构词。

英语中有相当数量的前缀和后缀能够使词义虚化，其中后缀数量最多、分布最广。常见的前缀有 pan-, inter-, trans-, pro-, ultra-, multi- 等。常见的后缀有：-ness, -tion, -ism, -sion, -ence, -ment, -ance, -ity, ship, -hood, -ing 等。这类词缀构成的抽象名词词义广泛，在英语里到处可见，例如：

Those virtues which characterize the young English gentlewoman, those accomplishments which become her birth and station, will not be found wanting in the amiable Miss Sedley, whose INDUSTRY and OBEDIENCE have endeared her to her instructors, and whose delightful sweetness of temper has charmed her AGED and her YOUTHFUL companions.

（2）用介词表达比较虚泛的意义。

介词在英语里非常活跃，可以构成各种各样的短语或成语，其意义有时甚至虚泛得难以捉摸。如：

in on 句式：

① I'd like to be in on the scheme.

我很想参与这项计划。

② Are you in on her secret?

你知道她的秘密吗？

be in for 句式：

① I am afraid we are in for a storm.

恐怕我们要赶上暴雨了。

② I'm in for the 800 meters.

我参加八百米赛跑。

3.1.3 悟性思维与理性思维

悟性思维也称直觉性思维，这种思维模式重视实践经验，注重整体思考，因而借助直觉和感觉从总体上模糊而直接地把握认识对象的内在本质和规律。理性思维注重科学、理性、分析、实证，必须借助逻辑，在论证、推演中认识事物的本质和规律。汉民族的思维方式重悟性，英美民族则重理性。

1. 汉民族的悟性思维

对中国人思维方式影响最大的三种哲学——儒家、道家与中国佛教都非常重视悟性。悟性思维对中国的语言文化影响普遍而深远，在文学、绘画、医学、宗教等方面皆有诸多表现，在语言上的表现如下：

（1）汉语重悟性的突出表现是意合。

意合的语言呈现出文学上的跳脱，特别是主语，代词和连接词常常略去不提，但是，意念流仍然大体清晰，跳脱部分，全凭"悟性"体味。如：

雨是最寻常的，（它）一下就是两三天，（不过）（你）可别恼。看，（它）（正在下着），（它）像花针（也）像细丝，（它）（那么）密密地斜织着，（以至于）屋顶上全笼着一层薄烟。

正如上例看到的，重直觉和悟性的思维必然导致语言的高度简约化。而语言的高度简约化又会反过来要求读者（听者）具有语言和直觉悟性，久而久之就形成了整个民族思维风格的传承。

（2）悟性思维导致汉语表述的广泛模糊化。

模糊化主要表现为：词性模糊化——汉语很多词的词性并不清楚，如"经济"可做名词也可作形容词；语义模糊化——比如"中央和地方"中的"地方"，界定就不明，到底是省，是市，还是一种统称；句法成分模糊化——汉语中常常难以确定主、谓、宾、动、状、补等句子成分，如"谁有闲工夫打听这件"这句话，到底是"有闲工夫"做"打听"的状语，还是"打听"做"有闲工夫"的补语，语法学家尚未给出定论；单复句界定的模糊化——如"我并没说什么，不过说了句顽话"这句是单句还是复句，语法学家们也是各持己见；动词形态的模糊化——如汉语动词没有时态、语态、语气这些形态标志。以上种种模糊化集中表现为汉语语法的隐含性。

（3）歇后语的运用也是悟性思维的表现。

中国民间的歇后语，通常由前段和后段组成，二者有意义上的联系，但在实际运用时，说话者常隐去后段，只说出前段，需要人从前段自己悟出后段，从而形成一种思维上的跳跃，这也是悟性思维的一种体现——要别人去顿悟。歇后语是汉语特有的现象，是中国智慧的产物，也是其他语言所没有的。

例如：

① 外甥打灯笼——照舅（旧）

② 猪八戒照镜子——里外不是人

2. 英美民族的理性思维

自从古希腊哲学家亚里士多德开创了形式逻辑以后，形式逻辑与理性主义就对英美民族的思维惯产生了深刻的影响。理性思维重逻辑理念，在语言上主要有以下表现：

（1）英语重形合是理性思维的表现。

形合就是用各种语言形式手段如形态变化、连接词等来表达语法和逻辑关系，它是英语表达法的一个重要特点，使得英语语言形式呈现出严谨的组织化程度。例如：

So if a man's wit be wondering, let him study the mathematics; for in demonstrations, if his wit be called away never so little, he must begin again。

（2）英语语法讲究精确性。

精确性是西方近代思维方式的一大特征。理性的严密推理往往从命题出发，英语基本句型中主语和谓语缺一不可，正是形式逻辑基本命题的需要。另外，英语的语法具有显性的特征，强调形式上的完整清晰，各种语法成分之间的关系必须通过形式准确地标定，如并列关系、从属关系、指代关系等。

例如：

① And then there was another Sunday and we were at Beon again that Sunday, and Russia came into the war and Poldand was smashed and I did not care about Poland, but it frightened France.

② The cook turned pale, and asked the housemaid to shut the door, who asked Brittles, who asked the tinker, who pretended.

从以上种种例句我们可以看出，英语是融理性思维与严谨的语言结构于一体的典范。

3.1.4 顺向思维与逆向思维

顺向思维是指按事物发展的先后顺序来进行推理、叙述或判断；逆向思维是指从截然不同甚至完全相反的角度来传达同样的信息。下面我们仅从几个方面对汉英民族顺向思维与逆向思维的差异做一简要介绍。

1. 时序表达上的后馈性与前瞻性

总体上说，汉民族由于历史悠久，重视回顾历史、尊重经验等，其思维传统是后馈性的，即以面向过去来区分时间上的前后，而英美民族则崇尚科学和理性，对自然和未来的发展富于好奇心，喜欢预测未来，提出假设、理论等，其思维方式是前瞻性的，即以面向未来来区分时间上的前后。这种截然相反的思维角度在语言中也有所体现。汉语用"前"指过去的时间，用"后"指未来的时间；而英语恰恰相反，用 back 指过去的时间，用 forward 指未来的时间。因此，汉语中有"前无古人，后无来者""前事不忘，后事之师""前赴后继""史无前例""前所未有"等说法；而英语中，这些"前"就变成了"后"，"后"也变成了"前"，如 But we are getting ahead of the story. (不过我们说到故事的后头去了)，这句英文就不能译为"不过我们说到故事的前头去了"。

2. 时间与空间排列顺序上的大与小

汉语在日期、钟点和地点上的表达是从大到小，而英语则习惯于从小到大。例如："大中小城市"在英语中是 small, medium-sized and large cities. 这一点在"整体思维与分析思维"部分已有论述，此处不再赘述。

3. 地理方位上的横与纵

在表达地理方位时，中国人习惯于先说横向方位名称、然后说纵向方位名称，即先"东西"后"南北"如东南、西北、东北、西南。

而英语人习惯于先纵向后横向，即 southeast, northwest, northwest, northeast。另外，在纵坐标轴上，"南北"的顺序也各不相同：中国人习惯于先南后北，如"南辕北辙""南征北战"，而英美人习惯于先北后南，如"转战南北"英语的说法是 fight north and south.

4. 叙事上的迂回与直接

在叙事上，汉民族习惯于从侧面说明、阐述外部环境，最后点出话语的信息中心；而英美民族往直截了当，把重要信息置于醒要位置。这表现在篇章结构上就是中国人写文章讲究起承转合，而英美人写文章往往开门见山，且章节脉络一目了然。表现在句式结构上就是汉语句式结构多为修饰成分较多，头长尾短，而英语多为句末重心，头短尾长。例如：

① 昨天上午八点半在博物馆门口我遇到多年未见的中学同学。

I met my middle school classmate at the entrance of the museum at 8∶30 yesterday morning whom I haven't seen for years.

② 昨晚在音乐会上，她钢琴弹得漂亮极了。

She played the piano beautifully at the concert yesterday evening.

5. 视觉思维倾向的差异

汉英民族思维顺序的差异也反映在认知世界，观察事物时所采取的不同视觉倾向、主要表现在以下几个方面：① 视角完全相反。如汉语中常说的"打八折"，英语常说 give somebody a 20 percent discount 或 20 percent off；汉语中的"抢险车"，英语为 a breakdown lorry，汉语中的"寒衣"，英语则为 warm clothes；汉语中的"消防队"，英语为 a fire brigade。② 视觉侧重点不同。这从一些事物的命名上有充分反映，如汉语的"挂钟"，英语叫 wall clock，汉语侧重方式，英语侧重地点；汉语的"隐形眼镜"，英语称为 contact lenses，汉语侧重外形，英语侧重方式；汉语的"戒指"，英语叫 finger ring，汉语侧重功能，英语侧重直观；汉语的"教学大楼"，英语为 classroom building，汉语强调的是用途，英语侧重组成部分；英语中叫做 lipstick 的东西,汉语叫"口红",汉语强调产品使用后的结果,而英语侧重其应用部位。观察着眼点不同。如中国人把金属分为"黑色金属"和"有色金属",相应的英语表述是 ferrous metal 和 nonferrous metal,前者着眼于金属表面呈现出的颜色,而后者着眼于金属的成分;汉语的"随手关门",英语常表达为 close the door behind you,汉语的表达着眼于手,而英语表达则着眼于人所处的位置;汉语的"仰卧""俯卧"是以面向为准,而英语表达方式 lie on one's back,lie on one's stomach 则以腹、背为准。

3.2 文化

何谓文化?关于文化的定义可谓多种多样。著名翻译理论家尤金·奈达曾将文化定义为"某一人群及其生存环境中所特有的各种活动、思想及其在物体和活动过程中所表现出来的物质形式的总和。"英国人类学家泰勒所下的定义是"一个复合的整体,其中包括知识、信仰、艺术、法律、道德、风俗以及作为社会成员而获得的任何其他的能力和习惯。"简言之,文化就是某一社会群体的整个生活方式(way of life)。语言与文化密不可分,相互作用。文化制约和影响语言的发展,语言是文化的载体,体现和反映文化。一位优秀的译者除了通晓两种语言文字外,还必须了解两种文化,深刻理解两种文化之间的差异。所以翻译者要注重培养和提高对文化差异的敏感性。下面简要叙述中西方文化差异及其在语言上的体现。

3.2.1 观念体系

由于文化形态不同,中西方在价值观念上存在很大差异,如 old 一词。"老"在汉语中表达的是尊敬的概念,如称某人为"张老"

"李老"是对其莫大的尊微。"老先生""老爷爷"等称呼也充分体现了中国人"尊老爱幼"的传统美德。然而,在英美人看来,old 是"不中用"的代名词,其意暗含老矣无能之意。英美人不喜欢别人说自己老,更不会倚老卖老。当有人对年纪大的外宾说"约翰先生,请坐。你年纪大,别累着"时,在中国人看来,这是对老人的一种关心,体现了尊老爱幼的美德,而如果把这句话直译为 Please sit down, Mr. John. You are old. Don't get tired。这位约翰先生会很不高兴,他会以为你暗示他风烛残年、老态龙钟。英美人不喜欢称他们为 grandma, grandpa,而更喜欢直呼其名,因为这类称呼与"精力、体力、能力下降"这一意义联系在一起。

在中西方文化中,宗教观念的差异体现非常明显。佛教在中国有近两千年的历史,佛教思想已深深植根于中国人的脑海里,对中国人的思维方式和语言表达产生了深远的影响。例如佛教认为求得真理的最好办法是静默、沉思等,之后真理会自然而然地显现,所以中国人的思维强调悟性,强调不言自明,不重视理性和逻辑推理,在语言上体现为语法的隐性,重意合不重形合。佛教语言也从方方面面渗入了中国人生活中:中国人生活中的词语许多跟佛教有关:刹那、解脱因缘、意识、觉悟等,还有一些俗语也来源于佛教,如:放下屠刀,立地成佛;苦海无边,回头是岸;恶有恶报,善有善报,等等。在英美等国家,人们信奉基督教。反映在思维方式上,就是西方人认为思想观念和现实世界之间存在着直接的联系,因此高度重视理性和逻辑,相信只要遵循正确的逻辑步骤就能求得真理。表现在语言上就是注重形式的严密,即语法上的形。基督教在英美人生活中的影响从其丰富的相关词汇、短语,表达方式中可见一斑,如 go to hell, bear one's cross, Man proposes, God disposes, God help those who help themselves 等。

3.2.2 自然条件

由于中国和英美国家在自然、环境、气候等方面的迥异，会带来汉英语言上的一些差异。中国地处北半球，来自太平洋的东风吹到地处东北信风带的中国，带来了一股暖流，是春季气候转暖的主要原因；而西风或西北风往往来源于西北高原，因此是凛冽的寒风。英国地处北温带，属海洋性气候，代表春天的到来的是西风。对于英国人来说，西风才是暖风，是生命的催化剂。英国处于高纬度，来自极地的东风异常寒冷，东风在英国是刺骨的寒风。由于地理位置的悬殊，西风和东风在两种文化中引发出不同的联想意义，汉语中有"万事俱备，只欠东风""等闲识得东风面，万紫千红总是春""东风汽车""东风百货"等。英国诗人雪莱的《西风颂》就是对春的赞美和企盼。"东风汽车"没有直接翻评成 east wind，而是译成 Areoles（风神），就是考虑到中西文化的差异。

在中国，普遍是河水东南流，所以有"一江春水向东流"等诗句，而英国河流大多向西北方向流入大海。所以李白的诗句"功名富贵若常在，汉水亦应西北流"只能译为"But sooner could flow backward to is fountains; This stream. than wealth and honor can remain."，译用 flow backward 来避免东西方因河流走向而引起的误解。

中国是一个内陆国家，幅员辽阔，地大物博，人们长期以来从事农业，几千年以来最发达的就是农耕文化，所以"牛"一词在汉语许多习惯用语中出现，如"牛脾气""牛刀小试""俯首甘为孺子牛"等。而英国是个岛国、海岸线很长，渔业比较发达，所以，英语 fish，water，sea 和许多表达方法有关系。如：a poor fish（可怜虫），to fish for fame and honours（沽名钓誉），all at sea（不知所措），keep one's head above water（奋力图存）等。当中国人用"牛饮"来描述喝酒特别多的情况，英国人则说 to drink like a fish；汉民族人们的生活离不开土地，固有"挥金如土"一说，而英语中则用 spend money like water 比喻花钱浪费，大手大脚。

3.3 习俗

习俗文化与日常生活和社交生活中的社会风俗、习惯紧密相连。世界上不同国家、不同民族有不同的风俗习惯和交际礼仪，风俗习惯往往反映一个国家或民族文化的外在特色。

3.3.1 称呼与称谓

在汉文化中，称呼与长幼、尊卑有关。幼小的、年轻的必须尊敬老的、年长的，地位低的必须尊敬地位高的。称呼比自己年长的人时，我们常常说"老李""李先生""张叔叔""王阿姨""何大妈"等。称呼比自己职位高的人要用其头衔加姓来称呼表示尊重，如"张局长""马院长""刘书记"等；在家庭成员之间，对长辈一定要称呼"爷爷""奶奶""爸爸""妈妈"等，绝对不能直呼其名。

在英语文化中，社交场合人们一般是用 Mr, Mrs, Miss, Ms, 加上姓氏来称呼，不管地位高低，如 Mr. Smith, Miss Brown 等；有时也可以用职业或头衔来称呼，如 President Bush, Professor Halliday, Dr. Nida 等，但是这种称呼应用范围有限，不像汉语中使用那么广泛；熟悉的人之间一般直呼其名，不仅平辈之间或者对同事、朋友如此，对于长辈也可以用名字直接称呼，甚至对老板和上级也可直呼其名。

3.3.2 问候

中国文化中，人们相遇打招呼喜欢说"吃饭了吗？""到哪里去?""上班去啊?""还在忙啊!""还没休息啊!""下班了?""刚回来?"等等。这些问候语自然地道,并没有什么特别的含义,只是打个招呼或引出话题而已,也表示一种互相关心。可以具体回答,也可以笼统地回答"办点事""出去趟"等。如果你回答"还没吃呢",对方也不一定要请你吃饭。

在英语文化中,问候语往往比较简单,如 hello., hi, good morning/ afternoon 等,回答相同;也有一些问候语是以 how 引导的特殊疑问句,如 How are you, How are things going 等,其功能也仅是打个招呼,因而总是期待肯定的答复,在比较正式的场合,可以用 It's very nice to meet you.

3.3.3 拜访

在中国,拜访一般不预先通知主人,原因大概是这样一来怕主人要花心思准备饭菜饮料等,会给主人造成负担,这是客人所不愿意的,也有的是想给主人来个惊喜,所以,不速之客多于预约的拜访。一般客来后都要泡茶招待,主人如果问客人想喝什么,往往得不到直接回答,这是因为中国的做客习惯是"客随主便"。斟茶时一定要双手端着茶杯送到客人手里才是礼貌的表现。喝茶的过程中主人会不断地为客人添水。客人不能把茶喝得精光,多少要剩一点,因为喝得精光是贪婪的表现,是不礼貌的。客人告辞时,主人多表示挽留,客人起身要走,主人要为其送行,并说一些客套话,如"请慢走""请走好""不远送啦""有时间再来"等,而客人则常说"请留步""不要送了""快请回吧"等,表示主人不必麻烦。

在英美国家,拜访他人首先要预约,与对方约好见面的时间、地点及内容。应邀到别人家里做客,到达要准时,不要早到也不要迟到(要注意不能到得太早,否则主人会觉得措手不及)。拜访时常给主人一点小礼物,如束花、一盒糖果等。主人家一般备有多种饮料,通常他们会问客人喝点什么,这时最符合西方礼仪又让主人高兴的方法就是做出具体的选择。在餐桌上,英美人常说 help yourself(请自便),而不是亲自动手给客人盛菜。告辞之前,要提前向主人暗示,可以说"I'm afraid I must be off now",并且对主人的招待表示感谢"It's been a wonderful evening.",有时第二天还要写张感谢便函或专门打电话感谢主人让你度过了一个非常愉快的夜晚。

3.3.4 致谢与答谢

致谢是文明社会的一种礼仪规范,是对他人帮助的承认,是促进人际关系的文明举动。汉文化中感谢语的使用因人际关系不同而各不相同。关系亲密的人之间几乎不用道谢,尤其是家庭成员之间,我们常听说"自己人,谢什么谢",因为汉民族人们之间关系越亲密客套越少,用了感谢语反而显得疏远和冷淡,是"见外"了。如果涉及公务往来和工作关系,如服务人员和服务对象的关系,纯属职责范围,并不属于帮助别人的情谊,则不必使用道谢语。英语文化中无论是家庭成员间,同事朋友间,还是上下级之间。不认识的人之间,只要别人为你做了一点事,提供了一点帮助,哪怕只是指指路,递个东西,你都必须道声谢,否则便是不礼貌。在英语文化中,"thank you"无处不在,人们事事都把它挂在嘴边,甚至上课时学生回答不出问题,老师还会说"谢谢"。英语常用的致谢语有:thank you,thank you very much, many thanks, a thousand thanks, It's very nice of you, I really don't know how to thank you enough,等等。

对对方的致谢必须做出回答,这就是答谢。汉语往往用否定形式,意在回绝,谦虚地表示不值一提,或者理应如此,用不着感谢。比如"甭客气""不用谢""不谢""没关系""不要紧",等等。而英语中常用的答谢语有:It's my pleasure; You are welcome; That's OK; Don't mention it 等。值得一提的是,汉语中的"没关系""不要紧""没事"相当于英语的 That's all right 或 That's OK，千万不能译成 It's doesn't matter 或 Never mind，因为这两句英语是对致歉语的回答用语。此外，汉语里的答谢用语"这是我应该做的"也不能想当然地译为 It's my duty 或 That's what I should do，因为在英美人看来，"职责""应该做的"给人的感觉是：你帮助我并非出于自愿，而是外界强加的责任而已，这正是英汉文化的差异所在。

3.3.5 称赞与回应

称赞是指对别人的赞美和欣赏，是一种交际活动和礼仪规范。汉语文化以谦逊为美德，中国礼貌制度的最大特点是"贬己尊人"。对自己或与自己相关的事务要"贬"，要"谦"，如自己的见解称"愚见"，自己的文章为"拙文"；对对方或与对方有关的事务要"褒"，要"赞"以示尊敬，如"贵国""贵公司""高见""大作""久仰大名"，等等。值得注意的是，称赞要分对象和场合，往往是上级对下级，父母对子女，长辈对晚辈称赞的多，反之，赞扬过多则有阿谀奉承之嫌。

在英语文化中，称赞有多种功能，人们在交际中十分重视称赞的作用，乐意听到别人对自己当面的称赞，也乐意称赞别人。赞美他人可以是对他人的肯定和鼓励，可以表达谢意，也可以作为打招呼的形式或引出话题的方式。英语中常用的称赞语有以下几种：① 名词短语 + is/look + 形容词，如：The chicken is great! Your dress looks so nice! 等；② 人称代词 + 动词 + 名词短语，如：I like your blouse. You did a good job! 等等；③ That's + 形容词 + 名词，如：That's a very beautiful skirt. 英语的称赞语中，80%以上都使用一些褒义的形容词，如 nice，good，beautiful，pretty，great，terrific，excellent 等。

对于称赞语的回答，汉英文化各不相同。汉语中对称赞语的反应往往是"拒绝 + 否定"的模式，不接受或不正面接受赞扬，这反映了中国人视谦虚为美德的传统。听到赞扬时,中国人常回答"哪里哪里""过奖了""不敢当""还差得远呢"等,表示受之有愧。有时若是女性被赞美长得漂亮等,往往还会觉得不好意思。而英语中对称赞语的回答常常是"接受+同意"的模式,表示对人的礼貌和尊重。通常会表示感谢,有时还会补充说明一下来历等,例如:

——This is such a fantastic flat! I really like it!

——Thank you, in fact, I spend two weeks decorating it.

3.4 历史

这里所说的历史是文化的历史发展与文化的历史沉淀在各自语言中的表现,尤指一些传统说法、成语典故、格言警句等。

汉语成语里蕴涵的特定文化意义非常丰富,如"四面楚歌""名落孙山""东施效颦""卧薪尝胆"等。下面分别来解释下它们的含义。

① 四面楚歌:项羽和刘邦原来约定以鸿沟东西边作为界限,互不侵犯。后来刘邦觉得应该趁项羽软弱的时候消灭他,就去追击项羽部队,布置了几层兵力,把项羽紧紧围在垓下,这时,项羽手下的兵士已经很少了,粮食又没有了。夜里听见四面围住他的军队都唱起楚地的民歌,不禁非常吃惊地说:"刘邦已经得到了楚地了吗?为什么他的部队里面楚人这么多呢?"说着说着,心里便已丧失了斗志,从床上爬起来,在营帐里面喝酒,并和他最宠爱的妃子虞姬一同唱歌。唱完一会儿,项羽骑上马,带了仅剩的八百名骑兵,从南突围逃走。边逃边打,到乌江畔自刎而死。因为这个故事里面项羽听见四周唱起楚歌,感觉吃惊,接着又失败自杀,所以后人就用"四面楚歌"这句话来形容人们遭受各方面攻击或逼迫的人事环境,而致陷于孤立窘迫的境地。

② 名落孙山：在宋朝的时候，有一个名叫孙山的才子和他同乡的儿子一同到京城，去参加举人的考试。放榜的时候，孙山的名字虽然被列在榜文的倒数第一名，但仍然是榜上有名，而和他一起去的那位同乡的儿子，却没有考上。不久，孙山先回到家里，同乡便来问他儿子有没有考取，孙山既不好意思直说，又不便隐瞒，于是，就随口念出两句不成诗的诗句来："解元尽处是孙山，贤郎更在孙山外。"从此，人们便根据这个故事，把投考学校或参加各种考试，没有被录取，叫做"名落孙山"。

③ 负荆请罪：战国时，蔺相如因多次为国争誉立功，被封为上卿，位于大将廉颇之上。廉颇心中不服，扬言如见到蔺相如就要羞辱他。蔺相如为顾全大局，多次退让躲避廉颇，致使廉颇深受动。于是廉颇便光着上身，身背荆条到蔺相如家请罪，从此两人结为生死之交，赵国亦将相和睦，国势大振。因此，"负荆请罪"就表示完全承认自己的错误，请求对方惩罚。

英语中的许多说法如 January chicks, green revolution, loneliness industry, white elephant, blue Monday, blue stocking, talk turkey, face the music 等都具有一定的历史文化内涵。

① January chicks：来源于乔叟的《坎特伯雷故事集》，是其中《商人的故事》中的一个主人公，名叫 January，它是 Lombard Baron 在 60 岁时与一位叫 May 的漂亮姑娘结婚后生的一子，所以 "to have January chicks" 指 "老来得子"。

② Green revolution：19 世纪末 20 世纪初，由于石油利用率大幅提高，农业生产中开始使用许多以石油为原料生产的化学产品，如农药、化肥、塑料薄膜等。这些石油化工产品广泛应用于农业，大大提高了农业生产力，这一现象后来被农业专家和历史学家称为绿色革命，即农业生产上的大革命、大革新。

③ Loneliness industry：美国由于现代工业文明的发展，越来越多的子女不与父母住在一起，因此 20 世纪六七十年代出现了大量的孤

独老人，他们无人照顾，生活艰难，成为一种社会问题。由此后来美国政府下定决心建立一种为孤寡老人服务的社会项目，叫做 loneliness industry，特指为孤独的人们服务的社会项目。例如：The United States has now set up a loneliness industry（美国现在已建立了一种为孤独老人服务的社会项目）。

④ White elephant：古罗马国王把白象送给自己不喜欢的朝臣作为一种惩罚，因为白象食量大，使主人不堪重负故。这里 white elephant 指庞大无用而累赘的东西。例如：A motor car would be a white elephant to him, because he can't drive（汽车对她来说毫无用处，因为她不会开车）。

⑤ Blue Monday：原指四旬节前的最后一个星期一，英美等国的人们习惯在这天开始痛饮，后来这个词语的意思发生了变化。人们度过轻松愉快的周末后，星期一上班工作时心情郁闷，20 世纪 50 年代多米诺的独唱歌曲 Blue Monday 风靡一时，使该词广为流传。

⑥ Blue stocking：在 18 世纪，伦敦有一个设在 Montagu 夫人家的俱乐部，男女成员相聚，以书刊评论和文化讨论代替无聊的闲谈，一反当时流行的吃喝玩乐、高谈阔论的风气。由于该俱乐部成员不穿绅士们常穿的时髦黑色长袜，而是穿普通蓝色长袜，被伦敦上流社会称为"蓝色俱乐部"，后来渐渐地人们就用 blue stocking 指"自视博学多才而貌不惊人的女子"。

典故多为形象生动的故事浓缩而成。英语中许多典故采自圣经和希腊罗马神话及古代寓言等，他们在英美文化中留下了深深的印记。比如 the apple of discord 来源于希腊神话：传说厄里斯女神未被邀请参加 Thetis 和 Peleus 的婚礼，她就把刻有"给最美丽的女人"的金苹果扔到参加婚礼的女神们中间以引起争端。结果特洛伊王子把苹果给了维纳斯，因为她许诺给王子天下最美丽的女人，这间接引起了古希腊人和特洛伊人之间的特洛伊战争。所以 the apple of discord 的意思是"争斗的原因或根源"。可见，Archille's heel 同样来源于希腊神话，传说 Archille 出生时被母亲握住脚跟倒浸在冥河中，因而除了脚跟外，他身体的其部分刀枪不入。但是在特洛伊战争中，它被特洛伊王子 Paris 用毒箭射中脚跟身亡。所以人们用 Archille's heel 比喻唯一致命的弱点。在英译汉时，熟悉英美文化历史传统有助于我们把握原文的隐含意义。

汉语里的很多俗语同样打着历史文化的烙印，如"月老"（介绍人）、"老泰山"（岳父）、"穿小鞋"（压制不同意见）、"戴高帽"（赞扬、奉承）、"抓小辫"（找岔子）。英语中一些俗语也与其历史文化紧密相关，如 Ark（避难所），Eden（乐园），Shylock（放高利贷者），baby-kisser（笼络人心的政客），kick the bucket（翘辫子），skeleton in the cupboard（不宜外扬的家丑），carry coals to Newcastle（徒劳无益，多此一举），meet one's Waterloo（遭到惨败），rob Peter to pay Paul（拆东墙补西墙）等。

3.5 社会心理

社会心理指由于各个民族的历史文化传统、生存形态、行为模式和交往原则等各不相同，因而形成了人们不同的社会心理习惯。现就英汉民族社会心理方面的几个问题比较如下：

3.5.1 集体主义与个体主义

中国人注重群体关系的和谐、群体目标的统帅和群体利益的维护。个体包含在整体之中，其核心是整体的利益，古代人提倡"修身、齐家、治国、平天下"，这里修身只是一种手段，目的是为了家、国与天下，可见从古人起就是以后者为重的。中国文化强调人与人之间的和谐相处，把整个社会当成一个"大家"，尊老爱幼、亲仁善邻、谦虚谨慎、互相关心，也有舍小家顾大家之说，譬如上古时期的大禹为了治水"过家门而不入"，备受赞誉。这种集体意识还体现在人们生活的方方面面：如人们见面打招呼问"哪里去啊？""吃了没？""干吗去呀？"都是表示一种亲密无间和关心；路遇不认识的人可以称呼"爷爷""大爷""大娘""大姐""大哥""小妹妹"等，因为在人们的潜意识中，大家都是"一家子"；人们闲谈之间常问及年龄、收入、婚姻状况等，即使是不熟悉的人谈论这些话题也不会觉得别扭，甚至有时还会热心地帮人介绍对象等；有意思的是男女之间谈恋爱被称为是"儿女私情"：这里的"私"显然与"公"相对，显示出它是自私的，不顾大局的。中国人的集体主义还反映在一种集体无意识和从众心理上，"不患寡而患不均"正是这种集体无意识的写照。另外汉语中很多习语和成语都含有对集体主义的颂扬，如："独木不成林""众人拾柴火焰高""三个臭皮匠，顶个诸葛亮""众志成城""大公无私""一枝独秀不是春"，等等。

西方人重视个体的独立，在西方文化中，个人是一个独立不依赖他人的自主实体。所以个体主义（individualism）的核心是强调每个人的价值必须按其本人的意愿与表现来对待和衡量。它不同于中国人心目中的"个人主义"，因为"个人主义"是"自私自利"的象征。在英美人看来，个体主义是一种积极的、奋发向上的精神，强调自我奋斗、自我控制、自我完善。在美国人的家庭中，孩子成年后他们有权利选择自己的生活，也承担由此而发生的所有相应的职责，他们18岁以后要离开父母独自居住，独立承担上大学的费用，即使是贷款也要自己还清。个体主义还强调个人的自由，这是个人自我实现的前提与条件，如言论自由、参加游行示威的自由、选择职业的自由等。由于个体主义强调个人，在考虑问题时往往突出个人利益，西方法律就非常注重对个人隐私权的保护。日常生活中约会一定要事先约好时间，突然造访是不会受到欢迎的；吃饭时如果事先没有说某人请客，就要各付各的账单；不能随意问人的年龄、收入、婚姻状况等隐私性的问题，除非对方主动提起。英语文化中的个人主义在语言上也有很多表现，如英语中有很多以 self 开头的词语来表达个人价值：self – awareness（自我意识）self – dependence（自立），self – fulfilling（自我实现），self – reliance（自我依靠），self – admiration（自我赏识）。

3.5.2 尊卑观念与平等观念

汉文化由于重视集体主义，所以注重秩序的概念，"国有国法，家有家规"，秩序、等级的观念深入人心，如尊卑有别，长幼有序等。这与中国封建社会持续了两千多年之久不无关系，正所谓"君君，臣臣，父父，子子"。在当代中国社会这仍然是需要注意的：如称呼长辈或者上级绝对不能直呼其名，只能用亲属称谓或者以其职位相称，如大伯、二哥、王局长，等等，否则便是对对方极大的不尊重；孩子起名字一般不用父母或祖辈的名字，否则是一种不敬。在家庭里，孩子不能顶撞父母，而父母教育孩子则天经地义。

基督教文化强调上帝面前人人平等，这就是西方平等精神最初的源泉。美国《独立宣言》指出：众人生而平等。这种平等的思想表现在生活的方方面面：学生在课堂上可以与教师辩论；人人可以批评政府或总统，甚至拿总统开玩笑；二十岁的年轻人可以对八十岁的老人直呼其名，对父母也是如此，等等。

汉文化中，人们遇到不相识的人需要互相介绍时，一般先介绍年纪大、身份比较高的长者，然后介绍年纪小、身份地位比较低的，这样才符合礼貌。而英语文化里介绍人认识时，首先提及妇女的名字才算有礼貌。例如：Mrs – Johnson, this is my colleague, Mr. Wolf.

3.5.3 数字观念

数字在各民族文化中都被赋予一些象征性的含义，在各自的文化心里中产生不同的联想。数字不仅是度量单位，古希腊哲学家毕达哥拉斯（Pythagoras）认为，数字是本原，是那种与所有东西保持内在联系的不变的结合物。在英汉文化里，都有一些吉祥数字和避讳数字。

在汉文化中，人们比较喜欢偶数，从"好事成双""四通八达"这类词语中可见一斑；客人来家里做客，菜的个数一定是偶数才好。其中"六"和"八"乃最吉祥之数。在汉文化里"六"象征着顺利，越多就越顺。在中国古代，民间称亲属为六亲，尊奉的神为六神，结婚男方向女方送六礼，音乐、舞蹈各六种，畜生也分六畜，等等。这说明"六"在汉文化里是一个极有分量的数。另外，"八"在中国文化里占有十分重要的地位，对八的崇拜很可能源于八卦。"八"是发达兴旺之数，因其谐音与"发"极其相似，168（谐音"一路发"），518（谐音"我要发"），888（谐音"发发发"）之类的号码备受青睐。值得一提的还有数字"九"，因为传说龙有九个儿子，它在中国传统文化中是与帝王有关的数字，如宫廷中的九龙柱、九龙壁等。现在，九仍是人们认为最为吉祥的数字。而提到数字"四"，由于谐音的关系，往往被认为是不吉利的数字，因为"四"和"死"发音极为接近。所以车牌或电话号码尾数如果含有四就很不受欢迎，尤其是14（谐音"要死"），514（谐音"我要死"），444（谐音"死死死"）。

在英语文化中，偶数被看作是阴性的、被动的，而奇数则被看作是阳性的、主动的。英美文化中，'three'是一个极为重要的数字，代表三位一体（the Trinity），基督教认为人由肉体、心灵和精神组成（body, mind and soul）；世界分为三部分：陆地、海洋和天空；"三"代表出生、生命和死亡；代表过去、现在和未来；代表男人、女人和儿童。甚至他们的政治制度也是三权分立的。数字"七"表示运动着的整个宇宙，在英美文化中是一个神奇而又吉利的数字，例如地球七大奇迹（seven wonders of the world），天使居住的七重天（the seven heavens in which the orders of angels dwell),人们常用的七连灯台（the seven branched candlestick）等，而且上帝在创世纪时第七天是休息日。今天，英美国家的人对七仍然非常重视，如美国制造的波音式飞机型号分别为707、727、737、747等。不过，"十三"在英美等国可是一个不吉利的数字，据说出卖耶稣的是他的第十三个门徒犹大（Juda），他为了贪图十三块银元而背叛了耶稣，传说魔鬼要有十二个女巫陪伴，因此共十三个成员去参加集会。西方人把十三看成大忌之数，重要活动不可安排在13日，共餐不可13人同桌，楼层、电梯、门牌号码等也都尽量避免这个数字。如果13日这天恰好赶上星期五，那更是倒霉透顶的日子，因为耶稣受难时正是星期五。

3.6 审美

世界各族人民在审美观念上往往存在很大差异：一个民族认为是美的事物在另一个民族看来可能并不是美的，甚至是丑的。这种差异是各民族不同的传统文化、民族心理、社会生活方式的体现。就汉语文化国家和英语文化国家来讲，各自的审美观也是大不相同。比如西方人虽是白种人，但喜欢晒太阳，把自己晒成浅麦色，以此为健康，以此为美。但中国人是黄色人种，却喜欢美白皮肤，有"一白遮百丑"之说，以肤白为美。再比如对待胖与瘦的态度。汉语中有"心宽体胖"之说，胖是有福气的象征，是"高贵相"。如果说某人胖了或发福了便意味着对他的赞美，说明生活过得比较顺心。人们见了小孩子也常说"胖嘟嘟""白白胖胖"等，表示对小孩子的喜爱；而英语中的fat则是一个含有贬义的词语，英美人对此讳莫如深，因为发胖常被看作是体质下降、愚笨的象征，用它来形容人非常不礼貌。《牛津高级英汉双解辞典》中对其用法的注释如下："fat is the most usual and direct adjective to describe people with excess flesh, but it is not polite…the most neutral term is overweight."所以英美人形容人胖多用 overweight。

3.6.1 关于动物

动物和鸟类在不同文化氛围里，其美感也各不相同。如"龙"与"凤","猫"与"狗","喜鹊"与"老鹰"等。

龙在汉文化传统中是勇敢、吉祥的化身，所以中国人自称为"龙的传人"，汉语中许多与龙有关的词语都是褒义的，如"龙飞凤舞""龙腾虎跃""望子成龙"等。而在英语文化中，龙（dragon）是一种可怕的怪物，因为它既像蛇又像鳄鱼，有翅有鳞，且能喷火吐烟。在开天辟地的神话里，龙是一种凶猛的原始动物，必须由神把它消灭。后来就有了许多骑士战胜恶龙的故事，从英国最早的史诗《贝奥伍夫》（Beowolf）对龙的描绘，我们可以大致领略西方龙的形象。由于这种差别，"亚洲小四龙"到英语里就变成了"four tigers"因为英语中 tiger 代表忠诚、勇敢和力量；而"拦路虎"则译成 a lion in the way，因为他们认为狮子才是百兽之王，是力量的象征。

在汉文化里，凤凰是传说中的一种神鸟，寓意"高贵""祥瑞""美丽"，如"龙凤呈祥"形容幸福美满，"丹凤朝阳"预兆稀世之瑞，"凤毛麟角"比喻稀世之珍，等等。在英语文化中，凤凰（phoenix）是一种不死鸟，每 500 年再生一次，它是复活、再生的象征，如：We all thought the airline was finished when it went bankrupt, but it rose like a phoenix from the ashes（当航空公司倒闭时，我们以为一切都完蛋了，然而它却像凤凰一样从失败的灰烬中获得了新生）。

汉语文化中狗的形象多为贬义，常被比喻为供人使役的牲畜或助人作恶的帮凶等，如："狼心狗肺""狗急跳墙""狗腿子""狗屁不通""狗眼看人低""狗嘴里吐不出象牙"等，甚至落水的狗也要痛打一顿——"痛打落水狗"。在英美国家，狗被视为人类最忠诚的朋友，如 lucky dog（幸运儿），top dog（最重要的人物），love me, love my dog（爱屋及乌），dog does not eat dog（同类不相残），every dog has his day（凡人皆有得意日），等等。

在汉文化里，猫是最得人缘的爱畜，也是民间的吉祥物。它还有一些别名如天子妃、财喜。现在有很多商家店铺都供有"招财猫"，象征富贵发财，这大概是源于宋元时期的谚语：猪来贫，狗来富，猫儿来，开质库（当铺）。在英语文化中，因为猫既温柔又狡诈，人们似乎对它有种不信任感。猫被看作是女巫的帮凶，常被比喻为坏女人、泼妇、长舌妇，甚至妓女。英语中猫的象征意义贬大于褒，如：a cat in the pan（叛徒），make a cat's paw of something（利用他人做工具），put the cat among the pigeons（制造麻烦）等。

喜鹊在汉文化中被视为喜庆的象征，人们出门做事如果听到喜鹊叫，就预示着能心想事成；喜鹊落在门前，预示喜事或贵客来临，人称"灵鹊报喜"；民间传说中还有喜鹊搭桥让牛郎织女相会的故事，因此喜鹊也就成为促成美满姻缘和预兆喜讯的象征。相反，在英美人的观念里，喜鹊（magpie）是一种聒噪的鸟，喜欢把各种漂亮的东西藏在自己的巢穴中，常被比喻为饶舌者和小偷，也象征嫉妒、自负和赶时髦。因此，在西方文化里，喜鹊的形象常常是阴郁的，它的出现往往预示着灾祸的降临，因而被看作是一种凶鸟。

在汉文化中，鹰的形象褒贬不一，虽然有时鹰可以象征英雄，但汉语中与鹰有关的成语俗语等多带贬义，如："鹰犬"比喻帮凶、爪牙；爪牙"鹰鼻鹞眼"形容奸诈凶狠之人的面貌；"鹰视狼步""鹰击毛挚"形容严酷、凶狠，等等，这大概与鹰的本性凶猛残暴不无关系。在西方文化里，鹰（eagle）是鸟中之王，是太阳、神和天神的象征，代表权威、力量、胜利和骄傲。在基督教中，鹰还是复活的基督的象征，鹰的高飞被比作基督的升天（Christ's Ascension）。由于鹰所具有的传奇般的美德，西方国家都把它视为自己的标志：美国把秃头鹰视为国鸟，采用秃鹰作为国徽，象征力量和自由；在英国，

eagle—badge 一直是英国皇家空军自豪的标志。

3.6.2 关于颜色

马克思曾说过，一般来讲色彩的感觉是美感的最普及的形式。红色在中国是喜庆的象征，结婚被称做"红喜事"，客人送礼叫做"送红包"，婚姻介绍人被称为"红娘"；"开门红"寓意从开始就取得好成绩，红色还象征革命，如"红军""红色政权""又红又专""一颗红心向着党"，等等。而在英美文化中，红色的联想意义大都与暴力、流血和战争有关，如 to see red 表示"火冒三丈"或"大发脾气"。因此，英国翻译家 David Hawkes 把《红楼梦》翻译为 A Story of the Stone 以避免 red 词在英美人头脑里产生的负面联想。汉语中的白色总会让人联想到死亡、鬼、披麻戴孝等，所以白色是种不吉利的颜色，汉语中丧事叫做"白事"，没有功名的人叫"白丁"，"白脸"象征奸诈，还有"白色恐怖"之说。在西方，白色象征纯洁和完美，婚礼上新娘要穿白色婚纱，婚礼马车要套白马，与白色有关的词语也多带褒义，如 days with a white stone（幸福的日子），white hat（好人），white lie（善意的谎言）等等。英国政府所在地叫做 White Hall，美国政府所在地叫做 White House。在中国。"黄色电影""黄色书刊""扫黄打非"之类的词总会让人联想到淫秽下流之意，而英语中的 yellow 一词却没有这样的联想意义，与汉语中"黄"相对应的是 blue，所以"黄色电影"应译为 blue film 而非 yellow film.

3.6.3 关于语言

从语言表达的审美观方面讲，汉语向来注重语音重叠及句法结构中的重复与对仗。语音重叠称为"叠词"，它便语流舒缓悠长，增强了汉语的表现力，在汉民族文化中极具审美价值。例如中国的古诗词中"寻寻觅觅,冷冷清清,凄凄惨惨戚戚""行行重行行,与君胜别离"之类的词句,给人以美妙的韵律感。而在英语中,只有极少数像 long long ago 之类的表达法用到了叠词。所以,像"绿油油""金灿灿"这一类的表达法就只能译为 green 和 golden,无法用类似的表达方式传递原文体现的美感。下面这则广告在汉语读者中会引起美妙的联想,而如果直译为英语恐怕会让英语读者感到不愉快:"十里蛙声不断,九溪曲流潺潺"。这会在汉语读者的脑海中构建出一个山水田园的美好图景,而英美人看了则会觉得这样的地方太吵,青蛙不停地叫,水流声又那么大,会打扰他们的安静。所以,懂得中西审美习惯的差异在翻译中非常重要,否则会导致翻译的失败。

简明建筑英语
翻译教程
Chapter 4

第四部分　翻译技巧及建筑文本翻译训练

4.1 翻译技巧：增益法和删减/省略法

建筑文本：中国建筑——故宫

外国建筑——法国凡尔赛宫

4.1.1 翻译技巧——增益法和删减/省略法

4.1.1.1 增益法（amplification）

增益法，是诸多翻译技巧中的一项，指在译文中增加一些原文字面上没有的词语。作为翻译的基本原则，译者是不允许随意增减原文意义的。但这并不是说译文和原文必须一词对一词。英语原文中一些隐含在字里行间的意义，因汉英表达方式不同，按字面照译无法说清的意思，在不失掉原文意义的前提下，为了使译文准确，均需要通过增词的手段来表达。

1. 语义因素

① We have always tried to educate our members to guard against self-complacency.

我们一直在教育我们的职员要防止自满情绪。（抽象名词后添加范畴词，使之符合汉语习惯）

② We should advocate the spirit of taking the whole situation into consideration. （添加"顾全大局"的实际所指含义）

要提倡顾全大局。

2. 语法因素

① And so we let the exciting new knowledge slip from us, a little further every day, and our confidence with it.

就这样，我们让那些振奋人心的新知识从身边悄悄溜走，日复

一日、渐行渐远，最后我们的信心也随之而逐渐丧失。（结构增补）

② Modesty helps one to go forward, whereas conceit makes one lag behind. （增补连词）

虚心使人进步，骄傲使人落后。

3. 修辞因素

① The crowd melt away.

人群渐渐散开了。（增添副词说明状态）

② It is high time that they should stop their argument. （强调）

他们早该停止争论了。

例句——汉译英

① 罗马竞技场是世界上规模最大的露天竞技场，外观呈椭圆形，周长超过五百米。

As the largest amphitheater in the world, it is a vast ellipse of over 500 meters in perimeter.

② 竞技场主要是用来进行角斗和举行其他公共活动。

The arena was designed mainly for gladiatorial contests and other public activities.

③ 比萨斜塔是比萨主教堂建筑群中四大建筑之一。

The Tower is one of the four buildings that make up the cathedral complex in Pisa.

④ 比萨斜塔是主教堂的独立钟楼，于1173年破土动工，因为战事频繁，工程持续了近两百年。

Grounded was broken in 1173, while construction continued for about 200 years due to the onset of a series of wars.

⑤ 体育馆设计新颖简约，成为2008年奥运会具有独特历史意义的地标性建筑。

The novel but simple design makes the Stadium a special historic landmark for the 2008 Olympic Games.

第四部分 翻译技巧及建筑文本翻译训练

例句——英译汉

① Originating in 12th - century France and lasting into the 16th century, it evolved from Romanesque architecture and was succeeded by Renaissance architecture.

哥特式建筑发源于 12 世纪法国，一直延续到 16 世纪。它由罗马式建筑发展而来，为文艺复兴建筑所继承。

② China is the cradle of an ancient and modern culture that has inspired the world with ideals.

中国是古代文化和现代文化的摇篮，这种文化以理想鼓舞着全世界。

③ In addition, the setting of the beam is simplified and coordinated with the line shape after the column optimization, which makes the structure of the tower more concise, more graceful and more powerful.

并且，塔柱优化以后，简化的横梁与线条相协调，这使铁塔结构更加简洁、优美、有力。

④ It is most familiar as the buildings of many of the great cathedrals, abbeys, and churches of Europe.

哥特式建筑常见于欧洲的大教堂、修道院和教堂建筑。

⑤ Although it was designed to be perfectly vertical, the tower started to lean during construction due to its adequate foundation set in a soft ground.

最初设计时塔是垂直竖立的，然而建造过程中，由于地基不均匀和图层松软，塔身开始倾斜。

⑥ The Taj Mahal was built by Mughal Emperor Shah Jahan in memory of his wife Mumtaz Mahal.

泰姬陵是莫卧儿皇帝沙·玛哈为纪念已故皇后慕塔芝·玛哈而建造的。

4.1.1.2 删减/省略法（omission）

译文中虽无其词，但其意通过上下文体现出来了，省略法使译文简明扼要，通顺易懂，符合目的语的行文规范。因此，与翻译的忠实原则并不背道而驰。

1. 英译汉中的省略法

——省略代词、冠词、连词、介词等。

① That's the way I am, and try as I might, I haven't been able to change it. （代词）

我就是这个脾气，虽几经努力，却未能改变。

② If winter comes, can spring be far away. （连词）

冬天来了，春天还会远吗？

③ The first electronic computer was produced in our country in 1958. （介词）

1958年我国生产了第一台电子计算机。

——省略动词。

当be，become，get，turn等系动词和形容词或介词连用，动词常常可以省略不译。

① When the pressure gets low, the boiling point becomes low.

气压低，沸点就低。

② Let's make an adjustment. （make，have等动词和具有动词含义的名词连用，可以省略不译）

我们调整一下吧。

——省略同义词或近义词。

① Neither party shall cancel the contract without sufficient cause or reason.

——省略赘语。

② MBA applicants who had worked at a job would receive preference over those who had not.

报考工商管理硕士的人,有工作经验的优先录取。

2. 汉译英中的省略法

——省略意义重复或重复出现的词语;

——省略原文中的范畴词;

——具体变抽象导致的成分省略;

——省略多余的描述成分;

——汉语中的排比、对偶,翻译时常省略冗余部分。

例如,

① 他感到了一种爱国热情在胸中激荡。

He felt the patriotism rise within his heart.

② 由于她突然出言不逊,我感到透不过气来。

Her unexpected rudeness made me gasp for breath.

例句——汉译英

① 苏州古典园林是中国江苏省苏州市园林建筑的总称。

The classical gardens of Suzhou are a group of ancient landscape gardens in Suzhou, Jiangsu province in China.

② 现存的胡同已被开发为一种不可多得的旅游资源。

The extant hutongs have been developed into rare sources of tourism.

③ 体育场总占地面积 20.4 万平方米,场内观众席约 91000 个,包括 80000 个固定座位和 11000 个临时座位。

Occupying an area of 204000 square meters, it has a seating capacity of 91000 with 80000 fixed seats and 11000 temporary seats.

④ 主体结构设计使用年限一百年,具备高度耐火、防水、抗震能力。

The main body of the Stadium has a design life of 100 years, with a high capacity of resistance to fire, water and the earthquake.

⑤ 这里经常举行祭孔仪式,参加祭祀的既有皇帝,也有钦命高官。

Sacrifices were often offered to the sage, either by emperors themselves, or by emperor-appointed high officials.

例句——英译汉

① John Roebling, the initial designer, was a great pioneer in the design of steel suspension bridges, who died just before construction.

初始的设计者约翰·罗布林是钢铁悬索桥的伟大先驱,然而在开工前他不幸去世。

② The most renowned Gothic architecture includes Cologne Cathedral in Germany, Westminster Abbey in Britain, and Notre-Dame de Paris in France.

最负盛名的哥特式建筑有德国的科隆大教堂、英国的威斯敏斯特教堂和法国的巴黎圣母院等。

③ The Colosseum is a vast elliptical amphitheater of ancient Rome. Made of concrete and stone, it is located east of the Roman Forum, in the center of Italy.

罗马竞技场是古罗马时期的一座巨大的椭圆形露天竞技场,位于意大利中心古罗马广场东面,以混凝土和石头筑成。

④ The Taj Mahal is widely recognized as one of the universally admired masterpieces of the world's heritage of great artistic value.

泰姬陵被普遍认为是世界遗产中最令人称道的杰作之一,具有极高的艺术价值。

⑤ The Brooklyn Bridge crosses majestically over New York City's East River, connecting the two broughs of Manhattan and Brooklyn.

布鲁克林大桥巍然横跨纽约东河,连接着曼哈顿岛和布鲁克林区。

4.1.2　建筑文本翻译实训

4.1.2.1　文本 A:中国建筑——故宫

(1)文本原文。

故宫

北京故宫,旧称紫禁城,位于北京中轴线的中心,是明清两个朝代的皇宫，是世界上现存规模最大、保存最为完整的木质结构的宫殿型建筑。故宫入选了世界文化遗产，是全国重点文物保护单位，国家AAAAA级旅游景区。

北京故宫于明成祖朱棣于西元1406年开始建设，明代永乐十八年（1420年）建成，曾有24位皇帝在此住过。故宫被誉为世界五大宫之首（北京故宫、法国凡尔赛宫、英国白金汉宫、美国白宫和俄罗斯克里姆林宫）。

故宫是一个巨大的建筑艺术瑰宝。建筑的艺术语言和表现手段非常丰富，包括空间、形体、比例、均衡、节奏、色彩、装饰等许多因素，正是它们共同构成了建筑艺术的造型美。空间，是建筑的基本形式要素，建筑主要通过创造各种内外空间来满足人们的实际需要，巧妙地处理空间，可以大大增强建筑艺术的表现力。

故宫的建筑艺术主要是群体组合的艺术，群体间的联系、过渡、转换，构成了丰富的铺陈展开的空间序列。故宫总体分为南部的前朝和北部的后寝两部分。南部以太和殿、中和殿、保和殿三大殿为中心，两侧辅以文华、武英两殿，是皇帝上朝接受朝贺、接见群臣和举行大型典礼的地方。三大殿建在高8米多的须弥座式三层平台上，四周环绕着石雕栏杆，气势磅礴，是故宫中最壮观的建筑群，表现出不同凡响的崇高地位。其中太和殿是故宫中等级最高，体量最大的建筑，也是我国现存的一座最大的古代木结构殿宇。北半部则以乾清宫、交泰殿、坤宁宫后三宫及东、西六宫和御花园为中心，其外东侧有奉先殿、皇极殿等，西侧有养心殿、雨花阁、慈宁宫等，是皇帝与后妃、皇子和公主们居住、举行祭祀和宗教活动以及处理日常政务的地方。

形体，主要指建筑物的总体轮廓。故宫的建筑气势恢弘，规模巨大。南北长961米，东西宽753米，占地72万多平方米，建筑总面积达16万多平方米，现存房屋8700余间。四周环绕高约10米的城墙和宽52米的护城河。城墙四周各设城门一座，南面午门，是故宫的正门，北面神武门，东面东华门，西面西华门。

　　故宫整个建筑空间变化丰富，体量雄伟，外观壮丽，有主有从，显示出庄严肃穆、唯帝王独尊的宏大气势。比例，主要是指巧妙处理建筑物各部分之间的比例关系，建筑中长宽高的比例，凹与凸的比例、虚与实的比例等，都直接影响到建筑美。

　　故宫整体建筑的比例和谐令人赞叹。以中国古代建筑外观上最显著的特征——屋顶形式为例，外朝三大殿（太和殿、中和殿与保和殿）的屋顶就各不相同。太和殿则是重檐庑殿顶，中和殿为四角攒尖顶，保和殿则是重檐歇山顶。

　　在不同屋顶形式的运用上，除了封建等级观念的影响外，也使这三座紧密相连的宫殿，在建筑形象上通过明显的对比而显得更加鲜明。尤其故宫的四座角楼，其屋顶结构更为复杂、奇巧，各部分比例谐调，檐角秀丽，造型玲珑别致，从而成为北京故宫的象征。均衡，主要指建筑在构图上的对称，包括建筑物前后、左右、上下各部分之间的关系。

　　均衡对称常常给人一种严肃庄重的感觉，增加崇高的美感。故宫作为一个完整的建筑群非常均衡对称，其中每座建筑物都是在一条由南到北的中轴线上展开，整个建筑群的中心是高大的太和殿，以此为中心由南向北伸展。

　　故宫宫殿建筑布局沿南北中轴线向东西两侧展开。节奏，指通过有规律的变化和排列，利用建筑物的墙、柱、门、窗等有秩序的重复出现，产生一种韵律美或节奏美，正是在这一点上，建筑和音乐具有同在的共同之处，因而人们把它们分别形容为凝固的音乐和流动的建筑。

　　我国著名建筑学家梁思成先生就曾经专门研究过故宫的廊柱，并从中发现了十分明显的节奏感与韵律感，从天安门经过端门到午门，就有着明显的节奏感，两旁的柱子有节奏地排列，形成连续不断的空间序列。

色彩，也常常构成建筑特有的艺术形象，给人们带来独特的审美感受和难忘的印象。北京故宫主要建筑的黄色琉璃瓦顶金碧辉煌、朱红色的柱子与门窗、檐下处于阴影部位的青绿色略点金的建筑彩画，在白色台基的衬托下，使建筑物各部分轮廓更加鲜明，色彩别具一格，从而使建筑物更加富丽堂皇。

故宫的建筑十分注意屋顶的装饰，不但在屋角处做出翘角飞檐，饰以各种雕刻彩绘，还常常在屋脊上增加华丽的走兽装饰。甚至故宫内各种门上九九排列的门钉，作为装饰也具有十分浓郁的民族文化内涵。

总之，正是通过空间、形体、比例、均衡、节奏、色彩、装饰等多种因素的协调统一，才形成了故宫建筑艺术特有的空间造型美。殿宇楼台，高低错落，壮观雄伟。故宫的建筑集中体现了中国古代建筑艺术的优秀传统和独特风格，是中国古代建筑的经典之作。

（编者整理后所得）

（2）参考译文。

The Palace Museum

The Imperial Palace in Beijing, formerly known as the Forbidden City and regarded as the Palace Museum presently, is located in the center of the central axis of Beijing. It is the palace of the two dynasties in the Ming and Qing Dynasties. It is the largest and most complete wooden structure in the world. The Imperial Palace has been selected as the world cultural heritage. It is a national key cultural relic protection unit and a national AAAAA grade tourist attraction.

Yongle, the founder of the Imperial Palace in Beijing, began construction in 1406. In the Ming Dynasty, Yongle was built in eighteen years (1420), and 24 emperors lived here. The Imperial Palace is known as the five largest palace in the world (the Imperial Palace in Beijing, Palace of

Versailles in France, the British chamber, the White House and Russia).

The Imperial Palace is a great treasure of architectural art. The art language and means of expression are very rich, including space, shape, proportion, balance, rhythm, color, decoration and many other factors, which constitute the beauty of the architectural art. Next we will appreciate the beauty of the architectural art of the Imperial Palace from these aspects. Space is the basic form factor of architecture. The building is mainly by creating all kinds of inner and outer space to meet the actual needs of people and skillfully dealing with space, which can greatly enhance the expressive force of the architectural art.

The architectural art of the Palace Museum is mainly the art of group composition. The connection, transition and transformation among groups form a rich series of spatial sequences. The Palace Museum is divided into two parts: the southern part for an emperor and officials' working and the northern part for his family activities. The three main hall of the southern part is Taihe Hall, Baohe Hall, and Zhonghe Hall, which are the halls of neutralization. The two sides of the halls are Wenhua Hall and Wuying Hall, where an emperor holds the large ceremonies, grants an interview to foreigners, and meets officials. The three central Halls are built on the three-story platform with more than eight meters high, surrounded by stone carving railings. It is the most magnificent building group in the Palace Museum, showing the lofty status of the extraordinary. Taihe Dian is the highest grade and largest volume building in the Imperial Palace. It is also the largest ancient wooden structure in China. In the north half, the Qianqing Palace, Jiaotai Hall, the Kun ning palace, the East and West Palace and the Imperial Garden are the center of the north half, and outside are the Fengxian hall, the Huangji hall. The western side includes the Yangxin hall, the Yuhua Pavilion, the Cining Palace and so on. They is the places where the emperor and the imperial concubines, princes and princesses live, hold the sacrificial and religious activities, and deal with the daily affairs.

Form, mainly refers to the overall outline of a building. The building of theImperial Palace is magnificent and huge. 961 meters long, 753 meters wide, covering more than 720 thousand square meters, with a total area of more than 160 thousand square meters and 8700 houses. It is surrounded by a wall of about 10 meters high and a moat 52 meters wide. The city walls are surrounded by gates, and the South Meridian Gate is the main entrance of the Imperial Palace, the northern gate of Shen Wu, the east gate of Donghua, and the west gate of Xihua.

The whole building space of the Imperial Palace is rich in space, magnificent in size and magnificent in appearance. It shows the grand momentum of solemn monarch. The proportion, mainly refers to the skillful handling of the proportion of the various parts of the building, the proportion of long and wide in the building, the proportion of concave and convex, and the proportion of virtual and real, all directly affect the beauty of the building.

The proportions of the overall building in theImperial Palace are very impressive. The roofs of the three main hall (Tai He Dian, Zhonghe Dian and Baohe Dian) are different from the most prominent features of the ancient Chinese architecture, the roof form. The hall of Taihe is the top of the double eaves and the hall of the hall. The hall of Central Harmony is a corner with four corners.

The use of different forms of the roof, in addition to the influence of the feudal hierarchy, also made the three closely connected palaces more distinct in the architectural image. Especially the four theImperial Palace buildings, its roof structure is more complex and ingenious, the proportion of each part is harmonious, the cornice angle is beautiful, the shape is exquisite, thus become the symbol of the the Imperial Palace in Beijing. Equilibrium refers to the symmetry of the building in composition, including the relationship between the front, left, top and bottom parts of the building.

Balanced symmetry often gives people a solemn feeling and adds sublime beauty. The the Imperial Palace, as a complete group of buildings, is very balanced and symmetrical, each of which is carried out on a central axis from south to north, and the center of the whole building is a tall temple of Taihe stretching from south to north.

The layout of theImperial Palace's palaces develops along the north - south axis to the East and west sides. Rhythm, refers to a regular change and arrangement, the use of the building's walls, columns, doors, windows and other orderly repeated appearance, producing a rhythmic beauty, on this point, architecture and music have the same common place, so they are described as solidified music and mobile buildings.

Mr. Liang Sicheng, a famous architect inChina, once specially studied the pillars of the Imperial Palace, and found a very obvious sense of rhythm. From the end door to the gate of the Tiananmen, there is a clear sense of rhythm. The columns on both sides are arranged rhythmically, forming a continuous space sequence.

Color often forms the unique artistic image of architecture, giving people unique aesthetic feeling and unforgettable impression. The yellow glazed tile roofs of the main buildings in Beijing the Imperial Palace are brilliant, vermilion columns and doors and windows, and the green and green color of the shade under the eaves is slightly gold. Under the white base of the table, the outline of the buildings is more distinct and the color is unique, thus making the building more magnificent.

The buildings of theImperial Palace pay much attention to the decoration of the roofs, not only in the corner of the roof, but also with various sculptures and ornaments on the roof of the house. Even the 99 door nails on various doors in the Imperial Palace have strong national cultural connotations as decoration.

In a word, it is through the coordination of many factors such as space, shape, proportion, balance, rhythm, color, decoration and so on, which has formed the unique beauty of space modeling in the architectural art of the Imperial Palace. The terrace of the palace, the high and low, magnificent. The architecture of the Imperial Palace embodies the fine tradition and unique style of Chinese ancient architectural art. It is a classic work of ancient Chinese.

（3）建筑专业术语及词汇。

① 皇宫/紫禁城 the Imperial Palace/ the Forbidden City/the Palace Museum

② 中轴线 central axis

③ 木质结构 wooden structure

④ 世界文化遗产 world cultural heritage

⑤ 国家 AAAAA 级旅游景区 national AAAAA grade tourist attraction

⑥ 重点文物保护单位 cultural relic protection unit

⑦ 内外空间 inner and outer space

⑧ 群体组合 group composition

⑨ 空间序列 spatial sequences

⑩ 石雕栏杆 stone carving railings

⑪ 总体轮廓 overall outline of a building

⑫ 均衡对称 balanced symmetry

⑬ 四角攒尖顶 Quadrangular pinnacle

⑭ 重檐庑殿顶 Double eaves veranda hall roof

⑮ 廊柱 pillar

⑯ 黄色琉璃瓦 yellow glazed tile

⑰ 高低错落 high and low dislocation

⑱ 门钉 door nail

⑲ 凹与凸 concave and convex

⑳ 翘角飞檐 corner eave

㉑ 朝代 dynasty

㉒ 皇帝，帝王 emperor

㉓ 瑰宝，宝藏 treasure

㉔ 节奏，韵律 rhythm

㉕ 装饰 decoration/ ornament

㉖ 构成，组成 constitute

㉗ 构图 composition of a picture

㉘ 廊柱 portico column

㉙ 台基 stylobate

㉚ 高耸的，巍峨的 lofty

㉛ 妾 concubines

㉜ 宏大气势 grand momentum

㉝ 庄严肃穆 solemn monarch

㉞ 突出的，杰出的 prominent

㉟ 华丽的，宏伟的 magnificent

㊱ 台阶，阶地 terrace

㊲ 过渡，转变，变迁 transition

㊳ 中立化，中立状态，中和 neutralization

㊴ 壕沟，护城河 moat

㊵ 封建等级制度 feudal hierarchy

㊶ 灵巧的 ingenious

㊷ 檐口 cornice

㊸ 庄严的，雄伟的 sublime

㊹ 布局，安排，设计 layout

㊺ 美的，美学的 aesthetic

㊻ 朱红色，鲜红色 vermilion

㊼ 民族文化内涵 national cultural connotation

㊽ 凝固的音乐 solidified music

㊾ 流动的建筑 mobile buildings

㊿ 经典之作 a classic work

4.1.2.2 文本 B：外国建筑——Versailles

（1）文本原文。

凡尔赛宫

位于巴黎西南18公里的凡尔赛宫，是法国最宏大、最豪华的皇宫。是人类艺术宝库中的一颗绚丽灿烂的明珠。

凡尔赛宫建于路易十四时代。他在位期间加强专制统治，强化中央集权。由于两次进行战争，使得晚年国库空虚，农民起义此伏彼起，法国封建专制制度开始走向没落。凡尔赛宫于1661年动土，1689年竣工，至今约有290年的历史。整个宫殿占地111万平方米，其中建筑面积为11万平方米，园林面积100万平方米。建筑气势磅礴，布局严密、协调。正宫东西走向，两端与南宫和北宫相衔接，形成对称的几何图案。宫顶建筑摒弃了巴洛克的圆顶和法国传统的尖顶建筑风格，采用了平顶形式，显得端正而雄浑。宫殿外壁上端，林立着大理石人物雕像，造型优美，栩栩如生。

凡尔赛宫宏伟、壮观，它的内部陈设和装潢富于艺术魅力。500多间大殿小厅处处金碧辉煌，豪华非凡。内部装饰，以雕刻、巨幅油画及挂毯为主，配有17、18世纪造型超绝、工艺精湛的家具。宫内还陈列着来自世界各地的珍贵艺术品，其中有远涉重洋的中国古代瓷器。由皇家大画家、装潢家勒勃兰和大建筑师孟沙尔合作建造的镜廊是凡尔赛官内的一大名胜。它全长72米，宽10米，高13米，连结两个大厅。长廊的一面是17扇朝花园开的巨大的拱形窗门，另一面镶嵌着与拱形窗对称的17面镜子，这些镜子由400多块镜片组成。镜廊拱形天花板上是勒勃兰的巨幅油画，挥洒淋漓，气势横溢，展现出一幅幅风起云涌的历史画面。漫步在镜廊内，碧澄的天空、静谧的园景映照在镜墙上，满目苍翠，仿佛置身在芳草如茵、佳木葱茏的园林中。

正官前面是一座风格独特的法兰西式大花园。园内树木花草的栽植别具匠心，景色优美恬静，令人心旷神怡。站在正官前极目远眺，玉带似的人工河上波光粼粼，两侧大树参天，郁郁葱葱，绿阴中女神雕塑亭亭而立。近处是两池碧波，沿池的铜雕塑丰姿多态，美不胜收。

凡尔赛官的修建，有一段历史轶事。1661年，居住在陈旧的凡赛纳官和枫丹白露官的路易十四，应财政总监大臣富盖邀请，去他新建的府第赴宴。富盖府第的富丽堂皇触怒了路易十四。三周之后，路易十四以贪污营私之罪将富盖投入监狱，并判处无期徒刑。嫉妒的心理促使路易十四做出兴建一座豪华皇官的计划。凡尔赛官的建造者，几乎全部是给富盖修建府第的人马，因此无论构造还是风格，两座建筑有异曲同工之妙。

历史上一度曾是法国政治、文化中心的凡尔赛在大革命后变得默默无闻了，到了19世纪下半叶，它又成为全世界瞩目的政治中心。1870年，普鲁士军队占领凡尔赛，第二年德皇在此举行加冕某些国家的君主即位时所举行的仪式，把皇冠戴在君主头上。同年，梯也尔政府盘踞在凡尔赛官，策划了镇压巴黎公社的血腥计划。此外，1873年，美国独立战争后，英美在此签订了《巴黎和约》。1919年6月28日，在镜廊里法国及英美等国同德国签订了《凡尔赛和约》，第一次世界大战宣告结束。

今日的凡尔赛官已是举世闻名的游览胜地，各国游人络绎不绝，参观人数每年达200多万，仅次于巴黎市中心的埃菲尔铁塔。南北官和正宫底层自路易·菲利浦起改为博物馆，收藏着大量珍贵的肖像画、雕塑、巨幅历史画以及其他艺术珍品。凡尔赛宫除供参观游览之外，法国总统和其他领导人常在此会见或宴请各国国家元首和外交使节。

（2）参考译文。

Versailles

In the 18km southwest of Paris, Versailles is the grandest, most luxurious palace in France. It is a bright pearl in the treasure-house of human art.

The palace of Versailles was built in Louis XIV. During his reign, he strengthened autocratic rules and strengthened centralization. After two wars, the old state Treasury was empty, the peasant uprising began, and the French feudal autocracy began to decline. In 1661 it ran a plant and it was completed in 1689. To this day it has a history of about 290 years. The whole palace covers an area of 1.11 million square meters. A building area was 110000 square meters, and a garden area was 1 million meters. Palace buildings are magnificent, with tight layout and coordination. The two ends of the palace are connected to the south and the north, forming a symmetrical geometric pattern. The palace roof abandoned the baroque domes, the French traditional steeple style, adopted the flat top form, appearing the correct and powerful. On the outside of the palace, there are statues of marble figures, beautiful and lifelike.

The palace of Versailles is magnificent and grand, of which interior furnishings and decoration are full of artistic charm. The hall of the palace, more than 500 rooms, is resplendent and magnificent. Interior decoration takes carving, large oil painting and tapestry and furniture made by superb craftsmanship, carrying with 17, 18th century modeling. The palace also contains valuable works of art from all over the world, including ancient

Chinese porcelain with a long history. The gallery, which was built by the royal great painter and decorator Le Blaine and the great architect Menschal is a great place of interest in the palace of Versailles. It is 72 meters long, 10 meters wide and 13 meters high, connecting two halls. On one side of the corridor are 17 large arched Windows facing the garden, and the other with 17 mirrors those are symmetrical with arched windows, which are made up of more than 400 lenses. On the vaulted ceiling of the gallery is a huge painting of Le Boland, dripping wet and full of vigor, showing a historical picture of the surging waves. Strolling in the mirror corridor, the clear sky and the quiet garden view mirror on the mirror wall. Everywhere is green, as if in the garden of green grass.

The front of the palace is a unique garden of French style. The trees and flowers in the garden show ingenuity. It is beautiful and tranquil landscape, and it is refreshing. Standing in front of the palace, a man – made river looks like the jade belt with its shimmering waves. The towering trees stand on both sides of the river, tall and green, and the goddess statue stands in the shade. Nearby are two pools of blue water, along which copper sculptures are abundant and beautiful.

The construction of the palace of Versailles has a historical anecdote that the world does not know much about. In 1661, Louis XIV, who lived in the archaic palace and Fontainebleau palace, was invited by the chancellor of the exchequer, Fulid, went to his new mansion. Louis XIV was offended by the magnificence of the Fugai's mansion. Three weeks later, Fulid was jailed for embezzlement and life imprisonment. The jealousy drove Louis xiv to build a grand palace. The builders of the palace of Versailles, almost all of the first to build the Fugai's mansion, so both the construction and the style of the two buildings had the same effect.

Once the political and cultural center of France, Versailles became an obscure country after the revolution. And in the second half of the 19th century, it became the political center of the world. In 1870, the Prussian army took possession of Versailles, and in the following year the German emperor held a ceremony to crown the throne of some countries and put the crown on the monarch's head. In the same year, the government of Thierry Henry was in the palace of Versailles, planning a bloody plan to suppress the Paris commune. Furthermore, in 1873, after the American Revolution, the United States and Britain signed the treaty of Paris. On June 28, 1919, the treaty of Versailles was signed between France and the United States and other countries in the mirror corridor, and the First World War ended.

Today, the palace of Versailles is a world – renowned tourist attraction, attracting more than 2 million visitors each year, second only to the Eiffel Tower in central Paris. The palace of the north and the south and the palace at the bottom changed to the museum from Louis Philippe, collecting priceless portraits, sculptures, giant historical paintings and other art treasures. In addition to the visit to the palace of Versailles, the French President and other leaders often meet or entertain heads of state and diplomatic envoys.

（材料摘自整理后所得）

(3) Terms and Vocabularies。

① geometric pattern 几何图案

② baroque domes 巴洛克的圆顶

③ French traditional steeple style 法国传统的尖顶

④ flat top form 平顶形式

⑤ statues of marble figures 大理石人物雕像

⑥ interior furnishings 内部陈设

⑦ interior decoration 内部装饰

⑧ carving 雕刻

⑨ the mirror gallery 镜廊

⑩ arched windows 拱形窗门

⑪ vaulted ceiling 拱形天花板

⑫ a unique French and western garden 一座风格独特的法兰西式大花园

⑬ man-made river 人工河

⑭ political and cultural center 政治文化中心

⑮ a world-renowned tourist attraction 举世闻名的游览胜地

⑯ tapestry 挂毯

⑰ artistic charm 艺术魅力

⑱ a great place of interest 名胜古迹

⑲ corridor 走廊

⑳ archaic palace 古代宫殿

㉑ reign 统治时期

㉒ centralization 中央集权

㉓ feudal autocracy 封建专制

㉔ resplendent 辉煌的，华丽的

㉕ vigor 活力，精力

㉖ stroll 散步，闲逛

㉗ ingenuity 心灵手巧，独创性

㉘ jade belt 玉带

㉙ shimmering 波光粼粼的

㉚ anecdote 轶事，奇闻

㉛ porcelain 瓷器

㉜ the Chancellor of the Exchequer 财政部大臣

㉝ mansion 宅邸

㉞ magnificence 壮丽，宏伟

㉟ embezzlement 盗用公款

㊱ life imprisonment 无期徒刑，终身监禁

㊲ obscure 黑暗的，模糊的

㊳ suppress 压迫

㊴ the Paris Commune 巴黎公社

㊵ the treaty of Paris 巴黎和约

㊶ portrait 肖像，画像

㊷ sculpture 雕塑

㊸ diplomatic envoy 外交使臣

㊹ peasant uprising 农民起义

㊺ coordination 协调

㊻ craftsmanship 手工艺

㊼ tranquil landscape 景色恬静

㊽ copper 铜制的

㊾ take possession of 占领

㊿ heads of state 国家元首

4.1.3 翻译强化训练

4.1.3.1 句子翻译

1. 汉译英训练题

① 北京故宫，旧称紫禁城，位于北京中轴线的中心。

② 故宫入选了世界文化遗产，是全国重点文物保护单位，国家AAAAA级旅游景区。

③ 故宫是一个巨大的建筑艺术瑰宝。

④ 建筑的艺术语言和表现手段非常丰富，包括空间、形体、比例、均衡、节奏、色彩、装饰等许多因素，正是它们共同构成了建筑艺术的造型美。

⑤ 空间，是建筑的基本形式要素，建筑主要通过创造各种内外空间来满足人们的实际需要，巧妙地处理空间，可以大大增强建筑艺术的表现力。

⑥ 南部以太和殿、中和殿、保和殿三大殿为中心，两侧辅以文华、武英两殿，是皇帝上朝接受朝贺、接见群臣和举行大型典礼的地方。

⑦ 它们是皇帝与后妃、皇子和公主们居住、举行祭祀和宗教活动以及处理日常政务的地方。

⑧ 南北长961米，东西宽753米，占地72万多平方米，建筑总面积达16万多平方米，现存房屋8700余间。

⑨ 故宫整个建筑空间变化丰富，体量雄伟，外观壮丽，有主有从，显示出庄严肃穆、唯帝王独尊的宏大气势。

⑩ 不同屋顶形式的运用，除了封建等级观念的影响外，也使这三座紧密相连的宫殿，在建筑形象上通过明显的对比而显得更加鲜明。

⑪ 故宫作为一个完整的建筑群非常均衡对称，其中每座建筑物都是在一条由南到北的中轴线上展开，整个建筑群的中心是高大的太和殿，以此为中心由南向北伸展。

⑫ 故宫宫殿建筑布局沿南北中轴线向东西两侧展开。

⑬ 我国著名建筑学家梁思成先生就曾经专门研究过故宫的廊柱，并从中发现了十分明显的节奏感与韵律感。

⑭ 两旁的柱子有节奏地排列，形成连续不断的空间序列。

⑮ 色彩，也常常构成建筑特有的艺术形象，给人们还来独特的审美感受和难忘的印象。

⑯ 在白色台基的衬托下，使建筑物各部分轮廓更加鲜明，色彩别具一格，从而使建筑物更加富丽堂皇。

⑰ 故宫的建筑十分注意屋顶的装饰，不但在屋角处做出翘角飞檐，饰以各种雕刻彩绘，还常常在屋脊上增加华丽的走兽装饰。

⑱ 甚至故宫内各种门上九九排列的门钉，作为装饰也具有十分浓郁的民族文化内涵。

⑲ 总之，正是通过空间、形体、比例、均衡、节奏、色彩、装饰等多种因素的协调统一，才形成了故宫建筑艺术特有的空间造型美。

⑳ 故宫的建筑集中体现了中国古代建筑艺术的优秀传统和独特风格，是中国古代建筑的经典之作。

2. 英译汉训练题

① It is a bright pearl in the treasure – house of human art.

② In 1661 it ran a plant and it was completed in 1689. To this day it has a history of about 290 years. The whole palace covers an area of 1.11 million square meters. A building area was 110000 square meters, and a garden area was 1 million square meters.

③ The palace of Versailles is magnificent and grand, of which interior furnishings and decoration are full of artistic charm.

④ The gallery, which was built by the royal great painter and decorator Le Blaine and the great architect Menschal is a great place of interest in thepalace of Versailles.

⑤ On one side of the corridor are 17 large arched Windows facing the garden, and the other with 17 mirrors those are symmetrical with arched windows, which are made up of more than 400 lenses.

⑥ The front of the palace is a uniquegarden of French style.

⑦ Nearby are two pools of blue water, along which copper sculptures are abundant and beautiful.

⑧ In 1661, Louis xiv, who lived in the archaic palace and Fontainebleau palace, was invited by the chancellor of the exchequer, Fulid, went to his new mansion.

⑨ In 1870, the Rrussian army took possession of Versailles, and in the following year the German emperor held a ceremony to crown the throne of some countries and put the crown on the monarch's head.

⑩ Today, thepalace of Versailles is a world – renowned tourist attraction, attracting more than 2 million visitors each year, second only to the Eiffel Tower in central Paris.

⑪ The palace of Versailles was built in Louis XIV. During his reign, he strengthened autocratic rules and strengthened centralization.

⑫ Palace buildings are magnificent, with tight layout and coordination.

⑬ The two ends of the palace are connected to the south and the north, forming a symmetrical geometric pattern.

⑭ The palace roof abandoned the baroque domes, the French traditional steeple style, adopted the flat top form, appearing the correct and powerful.

⑮ Interior decoration takes carving, large oil painting and tapestry and furniture made by superb craftsmanship, carrying with 17, 18th century modeling.

⑯ On the vaulted ceiling of the gallery is a huge painting of Le Boland, dripping wet and full of vigor, showing a historical picture of the surging waves.

⑰ Standing in front of the palace, a man – made river looks like the jade belt with its shimmering waves.

⑱ The construction of the palace of Versailles has a historical anecdote that the world does not know much about.

⑲ Louis XIV was offended by the magnificence of the Fugai's mansion. Three weeks later, Fulid was jailed for embezzlement and life imprisonment. The jealousy drove Louis xiv to build a grand palace.

⑳ The palace of the north and the south and the palace at the bottom changed to the museum from Louis Philippe, collecting priceless portraits, sculptures, giant historical paintings and other art treasures.

4.1.3.2 段落翻译

1. 汉译英训练题

① 平遥被称为"中国古典建筑的宝库",享有"明清时代建筑博物馆"的美誉。古城墙、镇国寺和双林寺被称为平遥古城"三宝"。从19世纪到20世纪初,平遥成为全国金融中心。现存的商铺和传统民居展现了古城昔日的繁荣昌盛。1997年平遥古城申遗成功,作为历史文化名城,平遥每年都吸引着无数游客。

② 颐和园位于北京市西北部海淀区,距市中心约15公里。颐和园是中国保存最完整的皇家园林,其著名的自然风光和文化意义对中国产生了重大影响,长期以来被誉为"皇家园林博物馆"。颐和园始建于1750年,原是一座豪华御苑,供皇室休闲娱乐之用,清朝末年成为皇室的主要居所。据历史文献记载,颐和园原名清漪园,1888年重建时才重新命名。颐和园完全展示出了自然之美和皇家园林的恢弘气势。景区主要由万寿山和昆明湖两部分组成,占地面积约三百公顷,共有人造古建筑三千多处,建筑面积七万多平方米。颐和园是中国首批国家级5A级风景区之一。1998年入选联合国教科文组织世界遗产名录。

③ 承德避暑山庄位于河北省承德市北部,是清代皇帝避暑和处理政务的场所,避暑山庄始建于1703年,历经康熙、雍正、乾隆三朝,历时89年建成。避暑山庄占地564万平方米,是中国现存最大的古典皇家园林。它与北京的颐和园、苏州的拙政园和留园并称中国四大名园。避暑山庄主要分为宫殿区和苑景区两部分。宫殿区位于山庄南部,是清代皇帝处理政务和居住之所。苑景区又分为湖泊区、平原区和山岳区,是清代皇帝休闲娱乐的场所。避暑山庄集中国古代造园艺术和建筑艺术之大成,是中国园林史上一个辉煌的里程碑和中国

古典园林艺术的杰作。1994年，承德避暑山庄及周围寺庙景区被列入世界遗产名录。

2. 英译汉训练题

① Baroque architecture is a building style originating in Italy in the 16th century, evolving from Italian Renaissance architecture, and lasting until the 18th century. The architectural style flourished in Italy by the middle of the 17th century and spread first to France and the throughout Europe. Baroque – style buildings share some common characteristics. Dramatic use of light is important in Baroque architecture, and is achieved either through strong light – and – shade contrasts or uniform lighting by means of windows. Opulent use of color and ornaments is prevalent, as can be seen in the large – scale frescoes painted on the ceilings. The most well – known Baroque buildings include the Saint Paul's Cathedral in UK and the Palace of Versailles in France.

② The Stonehenge is a mysterious prehistoric circle of upright stones in southern England. One of the most famous ancient sites in the world, the Stonehenge is composed of earthworks surrounding a circular setting of large standing stones. The current site as we can see today is only part of the original Stonehenge, since the original construction has suffered great damage from both weather and human being. The original purpose of building the Stonehenge remains unclear. But some people have speculated that it was a temple built for the worship of ancient earth deities. There is also archaeological evidence that it was once used as a burial ground from its earliest beginning. Whatever its purpose, the astonishing scale, the beauty of the stones, the skills of the construction, and the mystery surrounds it make the Stonehenge a famous site in England, attracting a large number of tourists each year.

③ Big Ben is the nickname for the Great Bell of the clock at the north end of the Palace of Westminster in London, and often extended to the Clock Tower. Located on the bank of Thames River, the tower is one of London's landmarks as well as one of the most famous Gothic architecture in the world. Nowadays the Tower has become one of the most prominent symbols of the United Kingdom. When a television or film maker wishes to indicate a generic location in the country, a popular way to do so is to show an image of the Tower. The Tower had its 150th anniversary in 2009, during which celebratory events took place. In 2012 the Clock Tower was officially renamed the Elizabeth Tower to celebrate the Diamond Jubilee of Elizabeth Ⅱ.

4.1.3.3 翻译强化训练参考答案

1. 句子翻译

（1）汉译英参考答案。

① The Imperial Palace in Beijing, formerly known as the Forbidden City and regarded as the Palace Museum presently, is located in the center of the central axis of Beijing.

② The Imperial Palace has been selected as the world cultural heritage. It is a national key cultural relic protection unit and a national AAAAA grade tourist attraction.

③ The Imperial Palace is a great treasure of architectural art.

④ The art language and means of expression are very rich, including space, shape, proportion, balance, rhythm, color, decoration and many other factors, which constitute the beauty of the architectural art.

⑤ The building is mainly by creating all kinds of inner and outer space to meet the actual needs of people and skillfully dealing with space, which can greatly enhance the expressive force of the architectural art.

⑥ The three main hall of the southern part is Taihe Hall, Baohe Hall, and Zhonghe Hall, which are the halls of neutralization. The two sides of the halls are Wenhua Hall and Wuying Hall, where an emperor holds the large ceremonies, grants an interview to foreigners, and meets officials.

⑦ They is the places where the emperor and the imperial concubines, princes and princesses live, hold the sacrificial and religious activities, and deal with the daily affairs.

⑧ North and South are 961 meters long, 753 meters wide, covering more than 720 thousand square meters, with a total area of more than 160 thousand square meters and 8700 houses.

⑨ The whole building space of the Imperial Palace is rich in space, magnificent in size and magnificent in appearance. It shows the grand momentum of solemn monarch.

⑩ The use of different forms of the roof, in addition to the influence of the feudal hierarchy, also made the three closely connected palaces more distinct in the architectural image.

⑪ The the Imperial Palace, as a complete group of buildings, is very balanced and symmetrical, each of which is carried out on a central axis from south to north, and the center of the whole building is a tall temple of Taihe stretching from south to north.

⑫ The layout of the Imperial Palace's palaces develops along the north – south axis to the East and west sides.

⑬ Mr. Liang Sicheng, a famous architect in China, once specially studied the pillars of the Imperial Palace, and found a very obvious sense of rhythm.

⑭ The columns on both sides are arranged rhythmically, forming a continuous space sequence.

⑮ Color often forms the unique artistic image of architecture, giving people unique aesthetic feeling and unforgettable impression.

⑯ Under the white base of the table, the outline of the buildings is more distinct and the color is unique, thus making the building more magnificent.

⑰ The buildings of the Imperial Palace pay much attention to the decoration of the roofs, not only in the corner of the roof, but also with various sculptures and ornaments on the roof of the house.

⑱ Even the 99 door nails on various doors in the Imperial Palace have strong national cultural connotations as decoration.

⑲ In a word, it is through the coordination of many factors such as space, shape, proportion, balance, rhythm, color, decoration and so on, which has formed the unique beauty of space modeling in the architectural art of the Imperial Palace.

⑳ The architecture of the Imperial Palace embodies the fine tradition and unique style of Chinese ancient architectural art. It is a classic work of ancient Chinese.

（2）英译汉参考答案。

① 它是人类艺术宝库中的一颗绚丽灿烂的明珠。

② 1661年动土，1689年竣工，至今约有290年的历史。全宫占地111万平方米，其中建筑面积为11万平方米，园林面积100万平方米。

③ 凡尔赛宫宏伟、壮观，它的内部陈设和装潢富于艺术魅力。

④ 由皇家大画家、装潢家勒勃兰和大建筑师孟沙尔合作建造的镜廊是凡尔赛宫内的一大名胜。

⑤ 长廊的一面是17扇朝花园开的巨大的拱形窗门，另一面镶嵌着与拱形窗对称的17面镜子，这些镜子由400多块镜片组成。

⑥ 正宫前面是一座风格独特的法兰西式大花园。

⑦ 近处是两池碧波，沿池的铜雕塑丰姿多态，美不胜收。

⑧ 1661年，居住在陈旧的凡赛纳宫和枫丹白露宫的路易十四，

应财政总监大臣富盖邀请，去他新建的府第赴宴。

⑨ 1870年，普鲁士军队占领凡尔赛，第二年德皇在此举行加冕某些国家的君主即位时所举行的仪式，把皇冠戴在君主头上。

⑩ 今日的凡尔赛宫已是举世闻名的游览胜地，各国游人络绎不绝，参观人数每年达二百多万，仅次于巴黎市中心的埃菲尔铁塔。

⑪ 凡尔赛宫建于路易十四时代。在位期间加强专制统治，强化中央集权。

⑫ 宫殿建筑气势磅礴，布局严密、协调。

⑬ 正宫东西走向，两端与南宫和北宫相衔接，形成对称的几何图案。

⑭ 宫顶建筑摒弃了巴洛克的圆顶和法国传统的尖顶建筑风格，采用了平顶形式，显得端正而雄浑。

⑮ 内部装饰，以雕刻、巨幅油画及挂毯为主，配有17、18世纪造型超绝、工艺精湛的家具。

⑯ 镜廊拱形天花板上是勒勃兰的巨幅油画，挥洒淋漓，气势横溢，展现出一幅幅风起云涌的历史画面。

⑰ 站在正宫前极目远眺，玉带似的人工河上波光粼粼，两侧大树参天，郁郁葱葱，绿阴中女神雕塑亭亭而立。

⑱ 凡尔赛宫的修建，有一段历史轶事。

⑲ 富盖府第的富丽堂皇触怒了路易十四。三周之后，路易十四以贪污营私之罪将富盖投入监狱，并判处无期徒刑。嫉妒的心理促使路易十四做出兴建一座豪华皇宫的计划。

⑳ 南北宫和正宫底层自路易·菲利浦起改为博物馆，收藏着大量珍贵的肖像画、雕塑、巨幅历史画以及其他艺术珍品。

2. 段落翻译

（1）汉译英参考答案。

① Pingyao is known as "a treasure house of Chinese classic architecture" and enjoys the reputation of "museum of architecture Ming and Qing dynasties". The ancient city walls Zhenguo Temple, and Shuanglin Temple are referred to as "Three Treasures of Pingyao". In the 19th and early 20th centuries Pingyao became the financial center of the country. The existing shops and traditional dwellings bear witness to its prosperity in the past. In 1997 the Ancient City of Pingyao was inscribed an the world Heritage List. As a famous historic and cultural city, Pingyao attracts numerous tourists every year.

② Situated in the Haidian District northwest of Beijing City, the Summer Palace is about 15 kilometers from central Beijing. As the best-preserved royal garden in China, it has greatly influenced Chinese horticulture with its famous natural views and cultural significance, and has long been recognized as "the museum of royal gardens". The construction started in 1750 as a luxuries royal garden for royal families to rest and entertain in. It later became the main residence of royal members in the end of the Qing Dynasty. According to historical documents, originally named Qingyi Garden (Garden of Clear Ripples), the Summer Palace was renamed after its reconstruction in 1888. The Summer Palace radiates fully the natural beauty and the grandeur of royal gardens. Composed mainly of Longevity Hill and Kunming Lake, it occupies an area of some 300 hectares. There are over 3000 man-made ancient structures, covering a floor space of more than 70000 square meters. The Summer Palace ranked among the first national 5A-level tourist spots in China and was inscribed on the UNESCO World Heritage List in 1998.

③ Located in the north of China, Hebei Province, the Chengde Mountain Resort was where Qing emperors took shelter from the summer heal and dealt with court affairs. Starting in 1703, the Resort was not completed until 89 years of construction during the resigns of the Qing Dynasty. Covering an area of 5.64 million square meters, the Resort mainly consists of two sections; the palace quarter and the scenery quarter. Located in the south of the Resort, the palace quarter was where the Qing emperors lived and dealt with state affairs. The scenery quarter, which is comprised of the lake area, plain area and mountain area, was where the emperors relaxed and entertained themselves. The Chengde Mountain Resort, an integration of garden and architecture construction techniques in ancient China, stands as a glorious monument in the Chinese history of gardens and a masterpiece of classic Chinese garden arts. In 1994, the Chengde Mountain Resort and its outlying temples were inscribed on the World Heritage List.

（2）英译汉参考答案。

① 巴洛克建筑源于16世纪的意大利，在意大利文艺复兴建筑的基础上发展而来，并延续到18世纪。这种建筑风格在17世纪中叶的意大利趋于繁荣，先是传入法国，随后传到欧洲各地。巴洛克风格的建筑有些共同的特点。一个重要特点是戏剧性地使用光线，通过营造强烈的光影对比或通过窗户的均匀照明来实现。艳丽的色彩和华丽的装饰十分普遍，这一点从天花板上的巨幅壁画中就能看得出来。最著名的巴洛克建筑有英国的圣保罗大教堂和法国的凡尔赛宫等。

② 巨石阵位于英格兰南部，是一个呈环形分布的/洋浦直立的石头组成的神秘史前石阵，巨石阵是世界著名古代遗迹之一，中间是环形的直立巨石，周围是土方结构。巨石阵最原始的结构经受了严重的岁月侵蚀和人为破坏，如今我们看到的巨石阵只是原始结构的一部分。建造巨石阵的初衷尚不得知。但是有些人推测，它是古时为供奉地球上的神灵而建造的庙宇。也有考古发现表明，巨石阵从建造初期就曾作为墓地使用。不管最初的建造目的是什么，巨石阵以其规模之大、巨石之美、建造工艺之高超、背景之神秘成为了英格兰的著名景点，每年吸引着大批游客。

③ 大本钟是伦敦威斯敏斯特宫北端钟楼的大报时钟的昵称，通常被用来指代整个钟楼。钟楼坐落在泰晤士河畔，是伦敦的标志性建

筑之一,也是世界上最著名的哥特式建筑之一。如今,钟楼已经成为英国的重要象征之一。电影或电视制作人通常会借用钟楼的形象,表示拍摄地是在英国某地。2009年为庆祝钟楼建成150周年,人们举行了隆重的庆典活动。2012年,为庆祝伊丽莎白二世登基60周年,大本钟所在的钟楼正式更名为"伊丽莎白塔"。

4.2 翻译技巧:重复法和正反译法

4.2.1 翻译技巧——重复法和正反译法

4.2.1.1 重复法

使语义清晰、连贯或为进一步加强语气,达到特定的修辞效果。汉语中的重复现象比英语多。

1. 英译汉中的重复法

(1) 重复英语中省略的成分(如名词、动词等);

(2) 重复英语中使用替代词表示的部分(如 nor, so, do, those, that 等);

(3) 重复同/近义词汇;

(4) 重复关系代词或关系副词;

(5) 重复叠字。

例子:

① Wood cannot conduct electricity, nor can glass.

木头不导电,玻璃也不导电。

② You will always find his tardiness and carelessness in everything he does.

你会发现他做任何事情都是磨磨蹭蹭,马马虎虎的。

2. 汉译英中重复结构的处理

（1）省略重复部分；

（2）替代重复部分；

（3）和并重复部分；

（4）保留重复部分。

例子：

① 特定文化，是特定社会政治和经济的反应，同时又对一定社会的政治和经济产生巨大的影响。

Any given culture is a reflection of the politics and economics of a given society, and the former in return has a tremendous influence and effect upon the latter.

② 在今后的五年内要实现经济状况的根本好转，实现社会风气的根本好转，实现党风的根本好转。

In the coming five years a fundamental turn for the better should be made in economic situations, in standards of social conduct and in Party style.

3. 例句——汉译英

① 自周秦以来，中国是一个封建社会，其政治是封建的政治，其经济是封建的经济。

From the Zhou and Qin dynasty onwards, Chinese society was feudal, as were its politics and economies. （省略）

② 中国人民历来是勇于探索、勇于创造、勇于革命的。

The Chinese people have always been courageous to probe into things, to make inventions and to make revolution.

③ 并且，塔柱优化以后，简化的横梁与线条相协调，这使铁塔结构更加简洁，优美，有力。

In addition, the setting of the beam is simplified and coordinated with the line shape after the column optimization, which makes the structure of the tower more concise, more graceful and more powerful.

④ 这是宝地！要不是宝地，怎么人越来越多？

This is a lucky place; if it isn't, why do more and more people come to live here?

4. 例句——英译汉

① We need materials which can bear high temperature and pressure.

我们需要能耐高温和耐高压的材料。

② China is the cradle of an ancient and modern culture that has inspired the world with ideals.

中国是古代文化和现代文化的摇篮，这种文化以理想鼓舞着全世界。

③ I very much admire your determination to strengthen China's economy and to ensure that the achievement are enjoyed by the masses who uphold China's socialist construction.

我非常钦佩你们决心增强中国经济，并且决心保证使从事社会主义建设的人民群众能够享受经济发展的成果。

④ In any case work does not include time, but power does.

在任何情况下，功不包括时间，但功率却包括时间。

⑤ Radar is a newly developed technique by which people can see the things beyond the visibility of them.

雷达是一种新开发的技术，利用雷达人们能看到视线以外的东西。

4.2.1.2 正反译法（Positive or Negative Translation）

在叙述同一件事情或者表达同一种思想时，英汉两种语言在表达习惯上有一定的差异。在英译汉时，英语里有些从正面表达的词或句子，译文中可从反面来表达。有些英语里从反面表达的词或句子，译成汉语时，需按汉语的表达习惯从正面来表达。总之，翻译时，我们应尽可能依照不同语言的习惯。

例句——英译汉

① Yet he was far from ready.

然而他远没准备好。（原文正面从表达，译文从反面表达）

② Don't lose time in posting this letter.

赶快把这封信寄出去。（原文从反面表达，译文从正面表达）

例句——汉译英

① 我不熟悉这种建筑风格。

I'm new to this architectural style.

② 可持续发展能力不断增强，生态环境得到改善。

The capability of sustainable development will be steadily enhanced; the ecological environment will be improved.

③ 博物馆内一切展品禁止触摸。

All the articles are untouchable in the museum.

④ 没有长城我们就无法感受到古代人的智慧与伟大。

We cannot feel the wisdom and greatness without the Great Wall.

⑤ 再聪明的工匠也难免出错。（智者千虑，必有一失）

It is a good workman that never blunders.

例句——英译汉

① The town center has changed beyond all recognition.

这个镇中心已经变得完全认不出了。

② Two inspections missed the fault in the engine that led to the crash.

两次检查都没发现引擎中导致事故的故障。

③ Our building materials are free from all the hazardous chemicals.

我们的建筑材料没有不含任何有害化学物质。

④ The failure of the international community to deal effectively with the problem has cost thousands of lives.

国际社会未能有效处理这个问题，导致了成千上万的人丧生。

⑤ I like that building; it's most unusual.

我喜欢那个建筑，它独具一格。

4.2.2 建筑文本翻译实训

4.2.2.1 文本 A：中国建筑——水立方

（1）文本原文。

水立方

基本介绍

国家水立方游泳中心又被称为"水立方"，位于北京奥林匹克公园内，是北京为2008年夏季奥运会修建的主游泳馆。国家游泳中心规划建设用62950平方米，总建筑面积65000~80000平方米，其中地下部分的建筑面积不少于15000平方米，长宽高分别为177m×177m×30m。

水立方与鸟巢分列于北京城市中轴线北端的两侧，共同形成相对完整的北京历史文化名城形象。奥运过后，水立方和鸟巢已成为北京市的新地标。

建筑特色

"水立方"位于奥林匹克公园B区西侧，和国家体育场"鸟巢"隔马路遥相呼应，建设规模约8万平方米，最引人注意的就是外围形似水泡的ETFE膜（乙烯-四氟乙烯共聚物）。ETFE膜是一种透明膜，能为场馆内带来更多的自然光。膜结构建筑是21世纪最具代表性的一种全新的建筑形式，至今已成为大跨度空间建筑的主要形式之一。它集建筑学、结构力学、精细化工、材料科学与计算机技术等为一体，建造出具有标志性的空间结构形式，它不仅体现出结构的力量美，还充分表现出建筑师的设想，享受大自然浪漫空间。

"水立方"是世界上最大的膜结构工程，除了地面之外，外表都采用了膜结构——ETFE材料，蓝色的表面出乎意料的柔软但又很充实。国家体育馆工程承包总经理谭晓春透露，这种材料的寿命为20多年，但实际会比这个长，人可以踩在上面行走，感觉特别棒。

"考虑到场馆的节能标准，膜结构具有较强的隔热功能；另外，修补这种结构非常方便，比如，射枪或者是尖锐的东西戳进去后，监控的电脑会自动显现出来。如果破了一个洞，只需用不干胶一贴就行了。膜结构还非常轻巧，并具有良好的自洁性，尘土不容易粘在上面，尘土也能随着雨水被排出。"谭晓春说，膜结构自身就具有排水和排污的功能以及去湿和防雾功能，尤其是防结露功能，对游泳运动尤其重要。

作为一个摹写水的建筑，水立方纷繁自由的结构形式，源自对规划体系巧妙而简单的变异，简洁纯净的体形谦虚地与宏伟的主场对话，不同气质的对比使各自的灵性得到趣味盎然的共生。椰树、沙滩、人造海浪……将奥林匹克的竞技场升华为世人心目中永远的水上乐园设计理念。

设计来源

这个看似简单的"方盒子"是中国传统文化和现代科技共同"搭建"而成的。中国人认为，没有规矩不成方圆，按照制定出来的规矩做事，就可以获得整体的和谐统一。在中国传统文化中，"天圆地方"的设计思想催生了"水立方"，它与圆形的"鸟巢"——国家体育场相互呼应，相得益彰。方形是中国古代城市建筑最基本的形态，它体现的是中国文化中以纲常伦理为代表的社会生活规则。而这个"方盒子"又能够最佳体现国家游泳中心的多功能要求，从而实现了传统文化与建筑功能的完善结合。

在中国文化里，水是一种重要的自然元素，并激发起人们欢乐的情绪。国家游泳中心赛后将成为北京最大的水上乐园，所以设计者针对各个年龄层次的人，探寻水可以提供的各种娱乐方式，开发出水的各种不同的用途,他们将这种设计理念称作"水立方"。希望它能激发人们的灵感和热情,丰富人们的生活,并为人们提供一个记忆的载体。

为达此目的,设计者将水的概念深化,不仅利用水的装饰作用,还利用其独特的微观结构。基于"泡沫"理论的设计灵感,他们为"方盒子"包裹上了一层建筑外皮,上面布满了酷似水分子结构的几何形状。水立方表面覆盖的 ETFE 膜又赋予了建筑冰晶状的外貌,使其具有独特的视觉效果和感受,轮廓和外观变得柔和。水的神韵在建筑中得到了完美的体现。轻灵的"水立方"能够夺魁,还在于它体现了诸多科技和环保特点。自然通风、循环水系统的合理开发,高科技建筑材料的广泛应用,都共同为国家游泳中心增添了更多的时代气息。

(材料编者整理后所得)

(2) 参考译文。

Water Cube

Brief Introduction

The National Water Cube swimming center, also known as the "Water Cube", is located in the Beijing Olympic Park, the main swimming pool built for the 2008 Summer Olympic Games in Beijing.

The National Swimming Center has 62950 square meters of planning and construction land, with a total building area of 65000 – 80000 square meters, of which the floor area of the underground part is not less than 15000 square meters, and the length and width are 177m × 177m × 30m respectively.

Water Cube and Bird's Nest stand on the two sides of Beijing's northern central axis, forming a relatively complete image of Beijing's historic and cultural city. After the Olympic Games, Water Cube and Bird's Nest have become the new landmark of Beijing.

Architectural features

"Water Cube" is located on the west side of the Olympic Park B area, and the National Stadium 'bird nest' across the road, the construction scale of about 80 thousand square meters, the most attractive is the peripheral ETFE film like bubble (ethylene Teflon). ETFE film is a transparent film, which can bring more natural light to the stadium. Membrane structure is the most representative form of architecture in the twenty – first Century. It has become one of the main forms of large – span space buildings. It combines architecture, structural mechanics, fine chemical, material science and computer technology to build a landmark space structure. It not only embodies the beauty of the structure, but also fully displays the architect's imagination and enjoys the romantic space of nature.

"Water Cube" is the world's largest membrane structure engineering, in addition to the surface, the appearance of the membrane structure of the TFE material, the blue surface is surprisingly soft but very substantial. Tan Xiaochun, the general manager of the construction contract of the national gymnasium, said the life of the material was 20 years, but it would be longer than that, and people could walk on it and feel great.

"In view of the Stadiums' energy saving standards, the membrane structure has a strong heat insulation function; in addition, it is very convenient to repair the structure, for example, when the gun or sharp objects are stamped in, the monitored computer will automatically appear. If a hole is broken, only a sticker is needed; the membrane structure is very light and has good self – cleaning. The dust is not easy to stick on it, and the dust can be discharged with the rain." Tan Xiaochun said, membrane structure itself has the function of drainage and sewage, and the function of dehumidification and anti – fogging, especially the anti – condensation function, is especially important for swimming.

As a building that depict water, the complex and free structure of the water cube derives from the ingenious and simple variation to the planning system, the simple and pure humility of the body and the grand home dialogue, and the contrast of different temperaments to their spirituality. Coconut, beach, artificial sea wave…The Olympic arena is sublimated into the eternal water paradise of the world.

Design Origin

This seemingly simple "square box" is built together by Chinese traditional culture and modern technology. The Chinese believe that nothing can be accomplished without norms and standards. In accordance with the established rules, we can achieve overall harmony and unity. In Chinese traditional culture, the "the circular heaven and the square earth" design thought hastened the "Water Cube". It echoed with the round "bird's nest", the National Stadium, and complemented each other. Square is the most basic form of ancient Chinese urban architecture. It embodies the rules of social life represented by outline ethics in Chinese culture. The "square box" can best reflect the multifunctional requirements of the National Swimming Center, thus realizing the perfect combination of traditional culture and architectural function.

In Chinese culture, water is an important naturalelement and arousing people's happiness. The National Swimming Center will be the largest water park in Beijing after the race, so the designer seeks the various forms of entertainment that water can provide for people of all ages and develop various uses of the water, which they call the "Water Cube". It is hoped that it can inspire people's inspiration and enthusiasm, enrich people's lives, and provide people with a memory carrier.

To achieve this goal, the designer deepened the concept of water, not only using the decorative function of water, but also using its unique microstructure. Based on the design inspiration of the "foam" theory, they wrapped a layer of architectural outer skins for the "square box", covered with the geometry of the cool water molecular structure. The ETFE film on the surface of the water cube also gives the appearance of ice crystal, giving it unique visual effects and feelings, and its contour and appearance are softer. The charm of water has been perfectly reflected in architecture. The "Water Cube" can win the championship in that it also embodies many technological and environmental characteristics. The rational development of natural ventilation, the rational development of circulating water system and the wide application of high tech building materials have added more times to the National Swimming Center.

(3) 建筑专业术语及词汇。

① 建筑特色 Architectural features

② 水立方 Water Cube

③ 膜结构 membrane structure

④ 建筑形式 form of architecture

⑤ 大跨度空间建筑 large – span space buildings

⑥ 建筑学 architecture

⑦ 结构力学 structural mechanics

⑧ 精细化工 fine chemical

⑨ 材料科学 material science

⑩ 节能标准 energy saving standards

⑪ 自洁性 self – cleaning

⑫ 排水和排污的功能 function of drainage and sewage

⑬ 去湿和防雾功能 function of dehumidification and anti – fogging

⑭ 防结露功能 anti – condensation function

⑮ ETFE 膜（乙烯 – 四氟乙烯共聚物）ETFE film (ethylene Teflon)

⑯ 透明膜 transparent film

⑰ 自然光 natural light

⑱ 对称排列 symmetrical arrangement

⑲ 视觉和声音的开始信号 visual and sound starting signals

⑳ 循环水系统 circulating water system

㉑ 体现；化身；具体化 embodiment

㉒ 代表；继任者；议员 representative

㉓ 大量的；结实的，牢固的；重大的；本质；重要材料 substantial

㉔ 隔离，孤立；绝缘；绝缘或隔热的材料隔声 insulation

㉕ 标记；用脚踩踏 stamp

㉖ 外围的；次要的；外部设备 peripheral

㉗ 透明的；清澈的；易识破的；显而易见的 transparent

㉘ 灵巧的；精巧的；设计独特的；有天才的，聪明的 ingenious

㉙ （使某物质）升华；使净化；纯化；升华物 sublimate

㉚ 永恒的，永久的；不朽的；永生的；永恒的事物 eternal

㉛ 天堂；伊甸园；乐园 paradise

㉜ 协调；融洽，一致；和谐；［音］和声 harmony

㉝ 统一体；统一性；团结一致 unity

㉞ 圆形的；环行的；流通的 circular

㉟ 多功能的，起多功能作用的 multifunctional

㊱ 几何学；几何形状；几何图形 geometry

㊲ 分子的；由分子组成的 molecular

㊳ 视觉的，看得见的；形象化的；画面，图象 visual

㊴ 空气流通；通风设备；通风方法 ventilation

㊵ 循环（的）；流通（的）circulating

㊶ 元素 element

㊷ 娱乐 entertainment

㊸ 载体 carrier

㊹ 泡沫 foam

㊺ 晶体；水晶 crystal

㊻ 轮廓 contour

㊼ 外观；外表 appearance

㊽ 环保的；环境的 environmental

㊾ 高科技 high tech

㊿ 合理的 rational

4.2.2.2 文本 B：外国建筑——The Arc de Triomphe

（1）文本原文。

凯旋门

凯旋门正如其名，是为庆祝胜利而创造出的纪念性建筑物，通常横跨在一条道路上单独建立，是一座迎接外出征战的军队凯旋的大门。

巴黎凯旋门是帝国风格的代表建筑。此种风格的崛起和拿破仑的倡导有着不可分割的关系。它的兴盛与衰败始终都与拿破仑的命运紧紧联系在一起，这些建筑都是以罗马帝国雄伟庄严的建筑为灵感和样板。它们尺度巨大，外形单纯，追求形象的雄伟、冷静和威严。巴黎凯旋门以古罗马凯旋门为范例，但其规模更为宏大，结构风格更为简洁。整座建筑除了檐部、墙身和墙基以外，不做任何大的分划，不用柱子，连扶壁柱也被免去，更没有线脚。凯旋门摒弃了罗马凯旋门的多个拱券造型，只设一个拱券，简洁庄严。

巴黎凯旋门，即雄狮凯旋门（Arc de triomphe de l'Étoile），是拿破仑为纪念1805年打败俄奥联军的胜利，于1806年下令修建而成的。拿破仑被推翻后，凯旋门工程中途辍止。波旁王朝被推翻后又重新复工，到1836年终于全部竣工。它是现今世界上最大的一座圆拱门，位于巴黎市中心戴高乐广场中央的环岛上面。这座广场也是配合雄狮凯旋门而修建的，因为凯旋门建成后，给交通带来了不便，于是就在19世纪中叶，环绕凯旋门一周修建了一个圆形广场及12条道路，每条道路都有40~80米宽，呈放射状，就像明星发出的灿烂光芒，因此这个广场又叫明星广场。凯旋门也被称为"星门"。1970年戴高乐将军逝世后，遂又改称为戴高乐将军广场。

巴黎凯旋门是欧洲100多座凯旋门中最大的一座，也是巴黎市四大代表建筑（即埃菲尔铁塔、凯旋门、卢浮宫和卢浮宫博物馆、巴黎圣母院）之一。巴黎凯旋门高约50米，宽约45米，厚约22米。四面各有一门，中心拱门宽14.6米。有凯旋门的四周都有门，门上有许多精美的雕刻。内壁刻的是跟随拿破仑远征的386名将军和96场胜战的名字和宣扬拿破仑赫赫战功的上百个胜利战役的浮雕。外墙上刻有取材于1792~1815年间法国战史的巨幅雕像。所有雕像各具特色，同门楣上花饰浮雕构成一个有机的整体，俨然是一件精美动人的艺术品。4组以战争为题材的大型浮雕："出征""胜利""和平"和"抵抗"；其中有些人物雕塑还高达五、六米。这其中最吸引人的是刻在右侧（面向田园大街）石柱上的"1792年志愿军出发远征"，即著名的《马赛曲》的浮雕，是在世界美术史上占有重要的一席之地的不朽艺术杰作。在这些巨型浮雕之上一共有6个平面浮雕，分别讲述了拿破仑时期法国的重要历史事件。凯旋门内部还刻有558位拿破仑帝国时代英雄的名字，其中一些人的名字下面划着线，那是因为这些人都是在战争中阵亡的。

凯旋门的拱门上可以乘电梯或登石梯上去，石梯共273级，上去后第一站有一个小型的历史博物馆，里面陈列着有关凯旋门的各种历史文物以及拿破仑生平事迹的图片和法国的各种勋章、奖章。另外，还有两间配有法语解说的电影放映室，专门放映一些反映巴黎历史变迁的资料片。再往上走，就到了凯旋门的顶部平台，从这里可以鸟瞰巴黎名胜。

凯旋门的正下方，是 1920 年 11 月 11 日建造的无名战士墓，墓是平的，地上嵌着红色的墓志："这里安息的是为国牺牲的法国军人"。据说，墓中长眠的是在第一次世界大战中牺牲的一位无名战士，他代表着在大战中死难的 150 万法国官兵。墓前有一长明灯，每天晚上，这里都会点起不灭的火焰。每逢节日，就有一面 10 多米长的法国国旗从拱门顶端直垂下来，在无名烈士墓上空招展飘扬。

巴黎 12 条大街都以凯旋门为中心，向四周放射，气势磅礴，为欧洲大城市的设计典范。

（2）参考译文。

The Arc de Triomphe

The Arc de Triomphe, as its name implies, is a memorial building created to celebrate victory. It is usually built separately across a road and serves as a gateway for the triumphant return of troops going out for battle.

The Arc de Triomphe in Paris is a representative building of the imperial style. The rise of this style has an inseparable relationship with Napoleon's advocacy. Its prosperity and decline have always been closely linked to Napoleon's fate. These buildings are inspired and modeled by the magnificent and solemn buildings of the Roman Empire. They are large in scale, simple in appearance, and pursue the majesty, calmness and dignity of the image. The Arc de Triomphe in Paris is exemplified by the Arc de Triomphe in ancient Rome, but its scale is larger and its structure style is more concise. Apart from the eaves, walls and wall foundations, the whole building does not make any big divisions, nor use pillars. Even buttress pillars are removed, and there is no architraves. The Arc de Triomphe abandoned the multi-arch model of the Roman Arc de Triomphe, setting up only one arch, which is simple and solemn.

The Arc de Triomphe de l'toile in Paris was built by Napoleon in 1806 to commemorate his victory over the Russian – Austrian coalition in 1805. After the overthrow of Napoleon, the Arc de Triomphe project stopped. After the overthrow of the Bourbon Dynasty, it was restored and finally completed in 1836. It is the world's largest circular arch, located in the center of Charles de Gaulle Square in central Paris on the roundabout. The square was also built in conjunction with the The Arc de Triomphe, which caused traffic inconvenience. In the mid – 19th century, a circular square and 12 roads were built around the Triumph Gate, each of which was 40 – 80 meters wide, making a radial shape, just like the brilliant light emitted by the stars. Therefore, the square is also called the Star Square. The Arc de Triomphe is also called the Stargate. After the death of General de Gaulle in 1970, it was renamed General de Gaulle Square.

The Arc de Triomphe in Paris is the largest of more than 100 triumphs gates in Europe. It is also one of the four representative buildings in Paris (the Eiffel Tower, the Arc de Triomphe, the Louvre and the Louvre Museum, Notre Dame de Paris). The Arc de Triomphe in Paris is about 50 meters high, 45 meters wide and 22 meters thick. There is one arch gate on each side, and the central arch is 14.6 meters wide. There are many exquisite carvings on the gates. Inside the gates are the names of 386 generals who followed Napoleon's expedition and hundreds of victorious battles that preached Napoleon's heroic achievements. The exterior wall is carved with giant statues of French history of war from 1792 to 1815. All the sculptures have their own characteristics. And with the flower relief on the lintel, they form an organic whole, forming a beautiful and moving work of art. There are four groups of large relief sculptures with the theme of war: "Expedition" "Victory" "Peace" and "Resistance". Some of them are as high as 56 meters. Among them, the most attractive one is the "1792 Volunteer Expedition" carved on the right stone pillar (facing the Rural Street).

The relief of the Expedition, the famous Marseille, is an immortal masterpiece of art which occupies an important place in the history of world art. On top of these huge reliefs, there are six planar reliefs, which respectively narrate the important historical events of France during Napoleon's period. Inside the Arc de Triomphe are 558 Napoleonic heroes, some of whose names are underlined because they were all killed in battle.

The arch of the Arc de Triomphe can be ascended by an elevator or by a stone staircase with 273 steps. At the first stop, there is a small museum of history, which displays various historical relics of the Arc de Triomphe, pictures of Napoleon's life and various medals of France. In addition, there are two film studios equipped with French explanations, which specialize in reflecting the historical changes of Paris. Further up, you can get a bird's eye view of Paris's scenic spots from the top platform of the Arc de Triomphe.

Right below the Arc de Triomphe is the Tomb of the Unknown Soldier, built on 11th, November 1920. The tomb is flat, with red epitaphs embedded on the ground: "Here lies the French soldiers who died for France." It is said that the tomb is buried for an unknown soldier who died in the First World War. He represents 1.5 million French officers and soldiers who died in the war. There is an altar lamp in front of the tomb. There is an everlasting flame every night. And on every festival, there is a French national flag, which is more than 10 meters long, hanging from the top of the arch, waving over the Tomb of the Unknown Martyrs.

The 12 streets of Paris are centered around the Arc de Triomphe, radiating around them with great momentum, which is a model for the design of large European cities.

(3) Terms and Vocabularies。

① the Arc de Triomphe 凯旋门

② memorial building 纪念性建筑物

③ triumphant return 凯旋而归

④ troop 军队

⑤ representative building 代表性建筑

⑥ commemorate victory 纪念胜利

⑦ the Roman Empire 罗马帝国

⑧ advocacy 拥护

⑨ prosperity and decline 兴盛和衰败

⑩ majesty 威严

⑪ dignity 尊严，高贵

⑫ exemplify 例证

⑬ the Bourbon Dynasty 波旁王朝

⑭ traffic inconvenience 交通不便

⑮ eave 屋檐

⑯ pillar 柱子

⑰ buttress pillar 扶壁柱

⑱ architrave 线脚

⑲ multi-arch model 多拱式

⑳ arch 拱门，拱券

㉑ circular arch 圆拱门

㉒ General 将军

㉓ Charles de Gaulle Square 戴高乐广场

㉔ roundabout 环岛

㉕ in conjunction with 连同；连接

㉖ preach 说教；讲道；鼓吹

㉗ radial shape 放射状

㉘ the Eiffel Tower 埃菲尔铁塔

㉙ the Louvre and the Louvre Museum 卢浮宫和卢浮宫博物馆

㉚ Notre Dame de Paris 巴黎圣母院

㉛ arch gate 拱门

㉜ carving 雕刻

㉝ sculpture 雕塑

㉞ expedition 远征

㉟ the Marseille 马赛曲

㊱ relief 浮雕

㊲ lintel 过梁，门楣

㊳ an organic whole 一个有机的整体

㊴ immortal masterpiece 不朽的杰作

㊵ stone staircase 石楼梯

㊶ top platform 顶部平台

㊷ historical relic 历史遗迹

㊸ a bird's eye view 鸟瞰

㊹ scenic spot 景点

㊺ epitaph 碑文

㊻ be embedded on 嵌入

㊼ altar lamp 长明灯

㊽ everlasting flame 永恒之火

㊾ the Tomb of the Unknown Martys 无名烈士之墓

㊿ momentum 势头

4.2.3 翻译强化训练

4.2.3.1 句子翻译

（1）汉译英训练题。

① 国家游泳中心（水立方）与国家体育场（俗称鸟巢）分列于北京城市中轴线北端的两侧，共同形成相对完整的北京历史文化名城

形象。

② 膜结构建筑是 21 世纪最具代表性的一种全新的建筑形式，不仅体现出结构的力量美，还充分表现出建筑师的设想，享受大自然浪漫空间。

③ 这个看似简单的"方盒子"是中国传统文化和现代科技共同"搭建"而成的。

④ 膜结构具有较强的隔热功能；另外，修补这种结构非常方便。

⑤ 膜结构还非常轻巧，并具有良好的自洁性，尘土不容易粘在上面，也能随雨水排出。

⑥ 膜结构自身就具有排水和排污的功能以及去湿和防雾功能，尤其是防结露功能，对游泳运动尤其重要。

⑦ 作为一个摹写水的建筑，水立方纷繁自由的结构形式，源自对规划体系巧妙而简单的变异，简洁纯净的体形谦虚地与宏伟的主场对话，不同气质的对比使各自的灵性得到趣味盎然的共生。

⑧ 将奥林匹克的竞技场升华为世人心目中永远的水上乐园设计理念。

⑨ 其内部是一个多层楼建筑，对称排列的大看台视野开阔，馆内乳白色的建筑与碧蓝的水池相映成趣。

⑩ 设计新颖，结构独特，融建筑设计与结构设计于一体。

⑪ 它不仅有使用和防卫的实用功能，在建筑艺术上也反映了封建礼制等级差异，也是建筑群体甚至园林成景的点缀。

⑫ 屋顶门扉上横竖各排列九颗镀金的圆钉，为殿宇式大门最高等级的做法。

⑬ 颐和园殿宇式大门的造型、装修、色彩都与各殿宇建筑群相配套，组成统一整体。

⑭ 垂花门在颐和园建筑中运用最多，造型最丰富也最富有特色。园内现存十余座，分布在前山前湖景区，有的作为园中之园出入口。

⑮ 有的作为园中之园出入口，有的串联于游廊和围墙之间，起

到障景和分割景区的作用。

⑯ 什锦门门框边饰以磨砖、砖雕，往往成为摄取景物的优美景框。

⑰ 什锦窗一般无格心，二层玻璃之间可以燃灯点缀园林夜景。除此之外，更多形式的窗户也是园林借景的重要手段。

⑱ 颐和园是中国现存古建筑规模最大、保存最完整的皇家园林，拥有七万平方米古建筑群落，包括宫殿、楼阁、亭桥、廊庑、轩榭、寺观、塔幢、城关、街市等。园林建筑是集各种建筑类型之大成。

⑲ 塔往往建于曲水转折处或山之巅峰，隐含镇守保平安的吉祥寓意。

⑳ 颐和园美的设计、美的传统让我们身心都能感受到古建园林之美。

（2）英译汉训练题。

① The Arc de Triomphe, as its name implies, is a memorial building created to celebrate victory. It is usually built separately across a road and serves as a gateway for the triumphant return of troops going out for battle.

② These buildings are inspired and modeled by the magnificent and solemn buildings of the Roman Empire.

③ They are large in scale, simple in appearance, and pursue the majesty, calmness and dignity of the image.

④ Apart from the eaves, walls and wall foundations, the whole building does not make any big divisions, nor use pillars.

⑤ The Arc de Triomphe abandoned the multi-arch model of the Roman Arc de Triomphe, setting up only one arch, which is simple and solemn.

⑥ The Arc de Triomphe de l'toile in Paris was built by Napoleon in 1806 to commemorate his victory over the Russian-Austrian coalition in

1805.

⑦ The square was also built in conjunction with the The Arc de Triomphe, which caused traffic inconvenience.

⑧ In the mid-19th century, a circular square and 12 roads were built around the Triumph Gate, each of which was 40-80 meters wide, making a radial shape, just like the brilliant light emitted by the stars.

⑨ The Arc de Triomphe in Paris is the largest of more than 100 triumphs gates in Europe. It is also one of the four representative buildings in Paris (the Eiffel Tower, the Arc de Triomphe, the Louvre and the Louvre Museum, Notre Dame de Paris).

⑩ There are many exquisite carvings on the gates. Inside the gates are the names of 386 generals who followed Napoleon's expedition and hundreds of victorious battles that preached Napoleon's heroic achievements.

⑪ There are four groups of large relief sculptures with the theme of war: "Expedition" "Victory" "Peace" and "Resistance".

⑫ The relief of the Expedition, the famous Marseille, is an immortal masterpiece of art which occupies an important place in the history of world art.

⑬ On top of these huge reliefs, there are six planar reliefs, which respectively narrate the important historical events of France during Napoleon's period.

⑭ Inside the Arc de Triomphe are 558 Napoleonic heroes, some of whose names are underlined because they were all killed in battle.

⑮ At the first stop, there is a small museum of history, which displays various historical relics of the Arc de Triomphe, pictures of Napoleon's life and various medals of France.

⑯ In addition, there are two film studios equipped with French explanations, which specialize in reflecting the historical changes of Paris.

⑰ Further up, you can get a bird's eye view of Paris's scenic spots from the top platform of the Arc de Triomphe.

⑱ It is said that the tomb is buried for an unknown soldier who died in the First World War. He represents 1.5 million French officers and soldiers who died in the war.

⑲ There is an everlasting flame every night. And on every festival, there is a French national flag, which is more than 10 meters long, hanging from the top of the arch, waving over the Tomb of the Unknown Martyrs.

⑳ The 12 streets of Paris are centered around the Arc de Triomphe, radiating around them with great momentum, which is a model for the design of large European cities.

4.2.3.2 段落翻译

(1) 汉译英训练题。

① 中国建筑是古代在遥远的东方发展起来的一种建筑风格,与西方建筑大不相同。中国传统建筑的主要特点是对称和平衡。引人注目的装饰性弯曲屋顶是中国建筑的一大特点。建筑风格在很大程度上依赖于木雕、石雕和石雕。中国建筑风格对日本、越南和韩国的建筑风格产生了巨大的影响,因此其总称是东方(东亚)建筑风格。

② 楼阁是体量较大的高层建筑。不仅是游人登高望远的佳处,更是园林中最为突出的景观。亭桥,"亭也,停也"。亭的功能主要是供游人作短暂的逗留以观览景色,也是园林中造景点景的重要手段,堪称中国古典园林中最具特色的建筑形式。亭的位置、式样、大小因地制宜,它们不仅可以停歇,其本身也是构成景色景点的重要组成部分。颐和园中现存各式各样亭子四十余座。它们或伫立于山岗之上、或依附于建筑之旁、或坐落于水池之畔、或掩映在花木之间,以

其美丽玲珑、丰富多彩的形象与园林中建筑、山水、植物相结合，构成一幅幅生动的画面。

③ 琉璃、彩画、砖雕、石雕、瓦当、露天陈设等作为建筑细节的一种装饰形式，体现着深厚的文化底蕴。瓦当具有保护木制屋檐和美化屋面轮廓的作用。琉璃用来增强建筑群庄严华贵的气氛，同时又具有优异的防水防腐功能。颐和园建筑彩画内容丰富、形式多样，尤以长廊 14000 余幅彩画最为突出，构成精美绝伦的艺术画廊。陈设包括建筑中的家具、字画、摆设灯具等所有器物，既能发挥园林建筑的使用功能，又起着装饰美化空间、营造意境的作用，并表达出时代的历史特征和园林主人的地位与品味。

（2）英译汉训练题。

① Fujian Tulou is a collective building built with earth walls. It is circular, semi-circular, square, quadrangular, pentagonal, chair-shaped, dustpan-shaped and so on. It has its own characteristics. Earlier earthen buildings were square, with palace style, mansion style and different postures. They were not only peculiar, but also mysterious and solid. Food, livestock and water wells are piled up in the building. If people need to resist the enemy, they only need to close the gate. Several young and middle-aged people guard the gate. The earth building is like a strong fortress. Women, children, old and young can rest easy. Because of the directionality of square earth buildings, the darkness of the four corners and the difference of ventilation and lighting, Hakka people have designed round earth buildings with good ventilation and lighting. Among the existing earthen buildings, circular buildings are the most noticeable, which are called round buildings or walled villages by local people.

② Fujian Tulou is a collective building. Its biggest feature is its large shape. Whether from a distance or in front of it, Tulou is shocked by its huge monolithic building. Its size is the largest of the residential buildings. The most common round building in the earth building is about 50 meters in diameter, three or four stories in height, with a total of more than 100 houses, which can accommodate three or forty families and two or three hundred people. The large round building, with a diameter of 780 meters and a height of 56 floors, has four or five hundred houses and can accommodate seven or eight hundred people. From the earthen building, this kind of dwelling style reflects the Hakka people's custom of living together. From the perspective of history and architecture, earthen buildings are built in a self-defensive way for the sake of ethnic security. Under the circumstances of Japanese invasions and civil wars, Hakkas who had migrated all over the country came thousands of miles to other places, choosing this kind of architecture which was conducive to family reunion and could defend against war. The descendants of the same ancestor form an independent society in an earthen building, coexisting and prospering, and sharing death and disgrace. Therefore, the imperial external condensation is probably the most appropriate summary of the earth building.

③ The walls of Fujian earth buildings are thicker under and thinner above, and some of them are 1.5 meters thick. When tamping, first dig deep and large trenches in the wall foundation, tamping solid, bury big stones as the foundation, and then use stones and mortar to build the wall foundation. Then they rammed the walls with splints. The raw materials of the earth wall are mainly local clayey laterite, which is mixed with appropriate amount of small stones and lime. After repeated mashing and mixing, it is made into what is commonly known as "ripe soil". Some key parts also need to be mixed with appropriate amount of glutinous rice, brown sugar to increase its viscosity. When tamping, Chinese fir branches or bamboo pieces should be embedded in the middle of the earth wall as "wall bone" to increase its tension. In this way, after repeated tamping, the earth wall like reinforced concrete was built, and a layer of anti-wind and rain erosion lime was put on the outside, so it is very strong and abnormal, and has good wind and earthquake resistance.

4.2.3.3 翻译强化训练参考答案

1. 句子翻译

（1）汉译英参考答案。

① Water Cube and Bird's Nest stand on the two sides of Beijing's northern central axis, forming a relatively complete image of Beijing's historic and cultural city.

② Membrane structure, the most representative form of architecture in the twenty-first Century, not only embodies the beauty of the structure, but also fully displays the architect's imagination and enjoys the romantic space of nature.

③ This seemingly simple "square box" is built together by Chinese traditional culture and modern technology.

④ The membrane structure has a strong heat insulation function; in addition, it is very convenient to repair the structure.

⑤ The membrane structure is very light and has good self-cleaning. The dust is not easy to stick on it, and can be discharged with the rain.

⑥ Membrane structure itself has the function of drainage and sewage, and the function of dehumidification and anti-fogging, especially the anti-condensation function, is especially important for swimming.

⑦ As a building that depict water, the complex and free structure of the water cube derives from the ingenious and simple variation to the planning system, the simple and pure humility of the body and the grand home dialogue, and the contrast of different temperaments to their spirituality.

⑧ The Olympic arena is sublimated into the eternal water paradise of the world.

⑨ Its interior is a multi-storey building with a wide view of large stands in symmetrical arrangement. The milk white building in the museum is very interesting to the blue water pool.

⑩ It has a novel design and unique structure, combining architectural design and structural design in one.

⑪ It not only has the practical function of use and defense, but also reflects the different level of feudal etiquette in architectural art.

⑫ Nine gold-plated round nails are arranged horizontally and vertically on the door leaf, which represents the highest level of the palace gate.

⑬ The shape, decoration and color of the palatial gates of the Summer Palace are matched with the palace buildings and form a unified whole.

⑭ They are in richest shapes and very distinctive. There are more than ten Chuihua Gates in the garden, which are distributed in the front hill and lake area.

⑮ Some of them are used as the entrance and exit of the garden, and some are connected in series between corridors and walls, which play the role of barriers and partition of the scenic spots.

⑯ The frames of this kind of gates are often decorated with rubbed bricks and brick carvings, which often become perfect frames for capturing a beautiful view.

⑰ Shijin windows have two layers of glass, through which lights can be lit to decorate the garden at night. In addition, more forms of windows are also important components of garden design as a whole.

⑱ The Summer Palace is the largest and most complete imperial garden in China. It has 70000 square meters of ancient building communities, including palaces, various pavilions, bridges, corridors, temples, towers, gateways and markets, composing a comprehensive collection of various architectural types.

⑲ Towers are often built at the turning point of winding waters or the peak of a hill, implying the auspicious meaning of keeping peace and tranquility.

⑳ The beautiful design and tradition of the Summer Palace make us feel the beauty of ancient gardens both physically and mentally.

（2）英译汉参考答案。

① 凯旋门正如其名，是为庆祝胜利而创造出的纪念性建筑物，通常横跨在一条道路上单独建立，是一座迎接外出征战的军队凯旋的大门。

② 这些建筑都是以罗马帝国雄伟庄严的建筑为灵感和样板。

③ 它们尺度巨大，外形单纯，追求形象的雄伟、冷静和威严。

④ 整座建筑除了檐部、墙身和墙基以外，不做任何大的分划，不用柱子。

⑤ 凯旋门摒弃了罗马凯旋门的多个拱券造型，只设一个拱券，简洁庄严。

⑥ 巴黎凯旋门，即雄狮凯旋门（Arc de triomphe de l'Étoile），是拿破仑为纪念1805年打败俄奥联军的胜利，于1806年下令修建而成的。

⑦ 这座广场也是配合雄狮凯旋门而修建的，因为凯旋门建成后，给交通带来了不便。

⑧ 于是就在19世纪中叶，环绕凯旋门一周修建了一个圆形广场及12条道路，每条道路都有40~80米宽，呈放射状，就像明星发出的灿烂光芒。

⑨ 巴黎凯旋门是欧洲100多座凯旋门中最大的一座，也是巴黎市四大代表建筑（即埃菲尔铁塔、凯旋门、卢浮宫和卢浮宫博物馆，

巴黎圣母院）之一。

⑩ 有凯旋门的四周都有门，门上有许多精美的雕刻。内壁刻的是跟随拿破仑远征的386名将军和96场胜战的名字和宣扬拿破仑赫赫战功的上百个胜利战役的浮雕。

⑪ 4组以战争为题材的大型浮雕："出征""胜利""和平"和"抵抗"。

⑫ "1792年志愿军出发远征"，即著名的《马赛曲》的浮雕，是在世界美术史上占有重要的一席之地的不朽艺术杰作。

⑬ 在这些巨型浮雕之上一共有六个平面浮雕，分别讲述了拿破仑时期法国的重要历史事件。

⑭ 凯旋门内部还刻有558位拿破仑帝国时代英雄的名字，其中一些人的名字下面划着线，那是因为这些人都是在战争中阵亡的。

⑮ 第一站有一个小型的历史博物馆，里面陈列着有关凯旋门的各种历史文物以及拿破仑生平事迹的图片和法国的各种勋章、奖章。

⑯ 另外，还有两间配有法语解说的电影放映室，专门放映一些反映巴黎历史变迁的资料片。

⑰ 再往上走，就到了凯旋门的顶部平台，从这里可以鸟瞰巴黎名胜。

⑱ 据说，墓中长眠的是在第一次世界大战中牺牲的一位无名战士，他代表着在大战中死难的150万法国官兵。

⑲ 墓前有一长明灯，每天晚上，这里都会点起不灭的火焰。每逢节日，就有一面10多米长的法国国旗从拱门顶端直垂下来，在无名烈士墓上空招展飘扬。

⑳ 巴黎12条大街都以凯旋门为中心，向四周放射，气势磅礴，为欧洲大城市的设计典范。

2. 段落翻译

（1）汉译英参考答案。

① Vastly different from Western architecture, Chinese architecture is a style of architecture that evolved in the far Orient in ancient times. The principal features of traditional Chinese architecture include symmetry and balance. Eye-catching, decorative, curved roofs are an instantly recognizable feature of Chinese architecture. The architectural style relies on timber work, stonework and stone carvings to a great extent. The Chinese style of architecture has greatly influenced architecture styles in Japan, Vietnam, and Korea and hence gets the overarching name: the Oriental (East Asian) style of architecture.

② A pavilion is a large high-rise building. It is not only the best place for visitors to climb high and see far away, but also the most prominent landscape in gardens. Ting, another pavilion, also has the same pronunciation with another Chinese character which means "stop", is a place for people to rest and enjoy the view for a short while. It is also an important part of scenic spots in gardens. Pavilions can be called the most distinctive architectural form in Chinese classical gardens. There are more than forty pavilions in Summer Palace. The location, style and size of pavilions are tailored to local condition. Some stand on hills; some are attached to buildings; some are situated on the banks of pools; while some are hidden between flowers and trees. With their exquisite and colorful images, pavilions are coordinated with garden buildings, landscapes, plants, all of which constitute a vivid and lovely picture.

③ As a decorative form of architectural details, colored glaze, colorful paintings, brick carvings, stone carvings, tiles, open-air display and so on all embody profound cultural connotations. The tile has the function of protecting the wooden eaves and beautifying the contour of the roof. Colored glaze is used not only to enhance the solemn and luxurious atmosphere of the architectural complex, but also has excellent functions of water-proof and anticorrosive. The colored paintings of the Summer Palace are rich in content and diverse in form, especially the more than 14000 colored paintings on the Long Corridor, which constitute an exquisite art gallery. The open-air display includes all the furniture, paintings, lamps and lanterns in the building, which are not only the functional components of garden architecture, but also play the role of decorating and beautifying the garden. They express the historical characteristics of times and create artistic conception to show the status and taste of the landscaper.

(2) 英译汉答案。

① 福建土楼是以土作墙而建造起来的集体建筑，呈圆形、半圆形、方形、四角形、五角形、交椅形、畚箕形等，各具特色。土楼最早时是方形，有宫殿式、府第式、体态不一，不但奇特，而且富于神秘感，坚实牢固。楼中堆积粮食、饲养牲畜、有水井。若需御敌，只需将大门一关，几名青壮年守护大门，土楼则像坚强的大堡垒，妇孺老幼尽可高枕无忧。由于方形土楼具有方向性、四角较阴暗，通风采光有别，所以客家人又设计出通风采光良好的圆土楼。现存的土楼中，以圆形的最引人注目，当地人称之为圆楼或圆寨。

② 福建土楼属于集体性建筑，其最大的特点在于其造型大，无论从远处还是走到跟前，土楼都以其庞大的单体式建筑令人震惊，其体积之大，堪称民居之最。在我们参观的土楼中最普通的圆楼，其直径大约为50余米，三、四层楼的高度，共有百余间住房，可住三、四十户人家，可容纳二三百人。而大型圆楼直径可达七八十米，高五六层，内有四五百间住房，可住七八百人。从土楼这种民居建筑方式体现了客家人聚族而居的民俗风情。从历史学及建筑学的研究来看，土楼的建筑方式是出于族群安全而采取的一种自卫式的居住样式。在当时外有倭寇入侵，内有年年内战的情势之下，举族迁移的客家人不远千里来到他乡，选择了这种既有利于家族团聚，又能防御战争的建筑方式。同一个祖先的子孙们在一幢土楼里形成一个独立的社会，共存共荣，共亡共辱。所以"御外凝内"大概是土楼最恰当的归纳。

③ 福建土楼的墙壁，下厚上薄，厚处有的竟达1.5米。夯筑时，先在墙基挖出又深又大的墙沟，夯实在，埋入大石为基，然后用石块和灰浆砌筑起墙基。接着就用夹墙板夯筑墙壁。土墙的原料以当地粘质红土为主，掺入适量的小石子和石灰，经反复捣碎，拌匀，做成俗称的"熟土"。一些关键部位还要掺入适量糯米饭、红糖，以增加其粘性。夯筑时，要往土墙中间埋入杉木枝条或竹片为"墙骨"，以增加其拉力。就这样，经过反复的夯筑，便筑起了有如钢筋混凝土般的土墙，再加上外面抹了一层防风雨剥蚀的石灰，因而坚固异常，具有良好的防风与抗震能力。

4.3 翻译技巧：状语从句的译法

建筑文本：中国建筑——鸟巢

外国建筑——法国埃菲尔铁塔

4.3.1 翻译技巧——状语从句的译法

英语状语从句包括表示时间、原因、条件、让步、目的等各种从句。常见的处理方法如下：

（1）表时间的状语从句。

——译成相应的表示时间的状语。

① While she spoke, the tears were running down.

她说话时，泪水直流。

② Please turn off the light when you leave the room.

离屋时请关电灯。

上句原文中表示时间的从句后置，译文中前置。

——译成"刚（一）……就……"的句式

① He had hardly rushed into the room when he shouted, "Fire! Fire!"

他刚跑进屋里就大声喊着:"着火了!着火了!"

② Hardly had we arrived when it began to rain.

我们一到就下雨了。

——译成并列的分句

① He shouted as he ran.

他一边跑,一边喊。

② They set him free when his ransom had not yet been paid.

他还没有交赎金,他们就把他释放了。

在上面两例的原文里,表时间的从句后置,在译文中提前。

(2) 表示原因的状语从句。

——译成表"因"的分句。

① The crops failed because the season was dry.

因为气候干旱,作物歉收。

② Because we are both prepared to proceed on the basis of equality and mutual respect, we meet at a moment when we can make peaceful cooperation a reality.

由于我们双方都准备在平等互尊的基础上行事,我们在这个时刻会晤就能够使和平合和成为现实。

"由于""因为"是汉语中常常用来表"因"的关联词,一般说来,汉语表"因"分句在表"果"分句之前,英语则比较灵活,但在现代汉语中,受西方语言的影响,也有放在后面的,如:

She could get away with anything, because she looked such a baby.

她能渡过任何风险,因为她看上去简直还像是个娃娃模样。

——译成因果偏正复句中的主句。

① Because he was convinced of the accuracy of this fact, he stuck to his opinion.

他深信这事正确可靠,因此坚持己见。

② Pure iron is not used in industry because it is too soft.

纯铁太软,所以不能用在工业上。

——译成不用关联词而因果关系内含的并列分句。

① "You took me because I was useful. There is no question of gratitude between us," said Rebecca.

"我有用,你才收留我。咱们之间谈不到感恩不感恩,"丽贝卡说。

② After all, it did not matter much, because in 24 hours, they were going to be free.

反正关系不大,二十四小时以后他们就要自由了。

(3) 表示条件状语从句。

——译成表"条件"的分句。

① "Sure, there's jobs. There is even Egbert's job if you want it."

当然,工作是有的。只要你肯干,甚至就可以顶埃格伯特的空缺。

② It was better in case they were captured.

要是把他们捉到了,那就更好了。

"只要""要是""如果""一旦"等都是汉语表示"条件"的常用关联词,在语气上,"只要"(或"只有")最强,"如果"最弱,"如果"也用来表示假设。英语中表示"条件"的从句前置后置比较灵活,汉语中表示"条件"的分句一般前置。

——译成表示"假设"的分句。

① If one of them collapsed, as they often did, the guide used to carry him over the mountains.

这种事时常在他们中间发生,如果其中一个人垮了,向导就要背着他过山。

② If the government survives the confidence vote, its next crucial test will come in a direct Bundestag vote on the treaties May 4th.

假使政府经过信任投票而保全下来的话,它的下一个决定性的考验将是在五月四日在联邦议院就条约举行的直接投票。

"如果""要是""假如"等都是汉语中用来表示"假设"的常用关联词,汉语中表示"假设"的分句一般前置,但作为补充说明情况的分句则往往后置。

——译成补充说明情况的分句。

① "He's dead on the job, Jess。Last night if you want to know."

"他是在干活时死的,杰西。就是昨晚的事,如果你想知道的话。"

② "You'll have some money by then, that is, if you last the week out, you fool."

"到那时你该有点钱了——就是说,如果你能度过这星期的话,你这傻瓜。"

(4) 表示让步的状语从句。

——译成表示"让步"的分句。

① Although he seems hearty and outgoing in public, Mr. Cooks is a withdrawn introverted man.

虽然库克斯先生在公共场合中是热情而开朗的,但他却是一个孤僻的、性格内向的人。

在上例中,两个分句的主语易位。

② While I grant his honesty. I suspect his memory.

虽然我对他的诚实没有异议,但我对他的记忆力却感到怀疑。

"虽然""尽管""即使""就算"等是汉语中表示"让步"的常用关联词。汉语中表示"让步"的分句一般前置(但现在也逐渐出现后置现象),英语中则比较灵活。

——译成表"无条件"的条件分句。

① Whatever combination of military and diplomatic action is taken, it is evident that he is having to tread an extremely delicate tight-rope.

不管他怎样同时采取军事和外交行动,他走的显然将不得不走一条极其脆弱的"钢丝"之路。

② No matter what misfortune befell him, he always squared his shoulders and said: "Never mind, I'll work harder."

不管他遭受什么不幸事儿，他总是把胸一挺，说："没关系，我再加把劲儿。"

汉语里有一种复句，即前一分句排除某一方面的一切条件，后一分句说出在任何条件下都会产生同样的结果，也就是说结果的产生没有什么条件限制。这样的复句里的前一分句，称之为"无条件"的条件分句。通常以"不论""不管""无论""管""随"等作为关联词。

（5）表示目的的状语从句。

——译成表示"目的"的前置分句。

① They stepped into a helicopter and flew high in the sky in order that they might have a bird's-eye view of the city.

为了鸟瞰这座城市，他们跨进直升飞机，凌空飞行。

② He pushed open the door gently and stole out of the room for fear that he should awake her.

为了不惊醒她，他轻轻推开房门，悄悄地溜了出去。

汉语里表"目的"的分句所常用的关联词有"为了""省（免）得""以免""以便""生怕"等等，"为了"往往用于前置分句，"省（免）得""以免""以便""使得""生怕"等一般用于后置分句。

——译成表示"目的"的后置分句。

① The murderer ran away as fast as he could, so that he might not be caught red-handed.

凶手尽快地跑开，以免当场被抓。

② Brackett groaned aloud, "You came from Kansas City in two weeks so that I could give you a job?"

布雷克特唉声叹气地说："你从堪萨斯城走了两个星期到这里，就是要我给你份工作吗？"

165

例句——汉译英

① 在中国南方建筑中,像在北方庭院建筑群中看到的被建筑包围的"开放空间"的概念并不常见,而南方建筑结构中可以看到的此类结构,称"天井"。

Although large open courtyards are less commonly found in southern Chinese architecture, the concept of an "open space" surrounded by buildings, which is seen in northern courtyard complexes, can be seen in the southern building structure known as the "sky well."

② 它们的整体框架是用砖或石头建造的,而门窗框架大多是用木头或铝合金制成的。

Their overall framework is built with bricks or stones, while the door and window frameworks are mostly made of wood or aluminum alloy.

③ 他们可以翻新并更容易地修复灾难造成的损害,避免搬迁,因此家庭成员必须学会生活在封闭的院落中,在他们整个生命中几乎没有个人空间,即使他们的角色随着年龄的变化而改变。

They could renovate and more easily repair the damage from disasters and avoid relocation, so family members had to learn to live in enclosed compounds with little personal space throughout their entire lives even as their roles changed as they aged.

④ 华北地区风大,冬天寒风从西北来,夏天风从东南来,北京四合院门开在南边,可保冬暖夏凉。

As cold wind in North China comes from the northwest, summer wind from the southeast, the gate of siheyuan (courtyard) opens in the south to keep warm in winter and cool in summer.

⑤ 在中国北方,如果一个建筑群有两层楼,第二层通常是一个阳台,作为保护主住宅免受攻击的建筑群北墙的一部分。

In northern China, if a compound had a two-story building, the second floor was usually a balcony set as part of the north wall of the compound that protected the main residence from attack.

例句——英译汉

① Although Buckingham Palace is furnished and decorated with priceless works of art that form part of the Royal Collection, one of the major art collections in the world today, it is not an art gallery and nor is it a museum.

尽管白金汉宫装饰着无价的艺术作品，构成了皇家收藏的一部分，并成为当今世界主要的艺术收藏之一，但它不是一个美术馆，也不是一个博物馆。

② While these buildings were revolutionary in their steel frames and height, their decoration was borrowed from neo – renaissance, Neo – Gothic and Beaux – Arts architecture.

虽然这些建筑在钢架和高度上都具革命性，但它们的装饰却借鉴了新文艺复兴时期、新哥特式和法兰西学院派艺术建筑。

③ While Gropius was active at the Bauhaus, Ludwig Mies van der Rohe led the modernist architectural movement in Berlin.

当格罗·皮乌斯积极创建包豪斯时，路德维·希密斯·范德洛领导了柏林的现代主义建筑运动。

④ After Constantine moved the capital of the Roman empire to Byzantium (now called Istanbul in Turkey) in A. D. 330, Roman architecture evolved into a graceful, classically – inspired style that used brick instead of stone, domed roofs, elaborate mosaics, and classical forms.

公元330年，君士坦丁将罗马帝国的首都迁至拜占庭（现土耳其伊斯坦布尔），罗马建筑演变成一种用砖代替石头，圆顶屋顶，精致的马赛克及经典形式的优雅的古典风格。

⑤ Even though architects are not given the same respect, influence and celebrity status as painters or sculptors, architecture is indeed one of the most progressive sciences for its continuous innovations and revolutionary developments that have evolved and rapidly enhance the human comfort and lifestyle.

尽管建筑师并不像画家或雕刻家那样得到同等的尊重、影响力，成为名人，但建筑因为其不断的创新和革命性的发展，并迅速提升了人类的舒适度并改善了生活方式，确实成为最进步的科学之一。

4.3.2 建筑文本翻译实训

4.3.2.1 Text A 中国建筑——国家体育场（鸟巢）

（1）文本原文。

国家体育场位于北京奥林匹克公园中心区南部，为2008年第29届奥林匹克运动会的主体育场。工程总占地面积21公顷，建筑面积258000平方米。场内观众坐席约为91000个，其中临时坐席约11000个。奥运会、残奥会开闭幕式、田径比赛及足球比赛决赛在这里举行。奥运会后这里将成为文化体育、健身购物、餐饮娱乐、旅游展览等综合性的大型场所，并成为具有地标性的体育建筑和奥运遗产。

国家体育场工程为特级体育建筑，主体结构设计使用年限100年，耐火等级为一级，抗震设防烈度8度，地下工程防水等级1级。工程主体建筑呈空间马鞍椭圆形，南北长333米，东西宽294米，高69米。主体钢结构形成整体的巨型空间马鞍形钢桁架编织式"鸟巢"结构，钢结构总用钢量为4.2万吨。混凝土看台分为上、中、下三层，看台混凝土结构为地下1层，地上7层的钢筋混凝土框架——剪力墙结构体系。钢结构与混凝土看台上部完全分开，互不相连，形式上呈相互围合，基础则坐在一个相连的基础底板上。国家体育场屋顶钢结构上覆盖了双层膜结构，即固定于钢结构上弦之间的透明的上层ETFE膜和固定于钢结构下弦之下及内环侧壁的半透明的下层PTFE声学吊顶。

国家体育场工程作为国家标志性建筑以及2008年奥运会主体育场,其结构特点十分显著且结构复杂。

设计理念

设计综述

国家体育场坐落于奥林匹克公园建筑群的中央位置,地势略微隆起,如同巨大的容器。高低起伏波动的基座缓和了容器的体量,而且给了它戏剧化的弧形外观。体育场的外观就是纯粹的结构,立面与结构是统一的。各个结构元素之间相互支撑,汇聚成网格状,如同一个由树枝编织成的鸟巢。在满足奥运会体育场所有的功能和技术要求的同时,设计上并没有被那些类似的过于强调建筑技术化的大跨度结构和数码屏幕所主宰。体育场的空间效果新颖前卫,但又简洁古朴,从而为2008年奥运会创造了独史无前例的地标性建筑。

基座

基座与体育场的几何体合二为一,如同树根与树。行人走在平缓的格网状石板步道上,步道延续了体育场的结构肌理。步道之间的空间为体育场来宾提供了服务设施:下沉的花园,石材铺装的广场,竹林、矿质般的山地景观,以及通向基座内部的开口。从城市的地面上缓缓隆起,几乎在不易察觉中形成了体育场的基座。体育场的入口处地面略微升高,因此,可以浏览到整个奥林匹克公园建筑群的全景。

屋顶

体育场的外观就是纯粹的结构,立面与结构是同一的。各个结构元素之间相互支撑,汇聚成网格状,就像编织一样,将建筑物的立面、楼梯、碗状看台和屋顶融合为一个整体。如同鸟会在它们树枝编织的鸟巢间加一些软充填物一般,为了使屋顶防水,体育场结构间的空隙将被透光的膜填充。由于所有的设施——餐厅、客房、商店和卫生间都是独自控制的单元,所以建筑外立面的整体封闭是非常不必要的。这使体育场可以做到自然通风,是体育场环保设计的最重要的一个方面。

碗状看台

体育场被设计成为巨大的人群的容器，无论远眺还是近观，都给人留下与众不同的、不可磨灭的印象。体育场内部，这种均匀的碗状结构形体将能调动观众的兴奋情绪，有可能使运动员超水平发挥。创造连贯一致的外表，座位的干扰被控制到最小，声学吊顶将结构遮掩使得观众和场地上的活动成为注意焦点。在此，人群形成了建筑。

（2）参考译文。

Located at the southern part of the Olympic Park in Beijing, the National Stadium is the main stadium of the 29th Olympic Games in 2008. Occupying an area of 21 hectares, it has a floor space of 258000 square meters. Its seating capacity amounts to 91000, including 11000 temporary seats.

The venue hosted the opening and closing ceremonies of the Beijing Olympic Games and Paralympic Games, the track and field competitions, and the football finals. After the Olympics, the stadium has become a large-scale sports and entertainment facility for the residents of Beijing—an architectural landmark and Olympic legacy.

The main body of the National Stadium has a design life of 100 years. Its fire resistance capability is first-rate, and it can withstand an eight-magnitude earthquake. The water-resistance capability of its underground project is also first-rate.

The main body of the National Stadium is a colossal saddle-shaped elliptic steel structure, weighing 42000 tons. It is 333 meters long from north to south, 294 meters wide from east to west, and 69 meters tall. The main body's elements support each other and converge into a grid formation, just like a bird's nest with interlocking branches and twigs. Being a seven-story shear wall system, the stadium's stand has a concrete framework. The upper part of the stand and the stadium's steel structure are separated from each other, but both are based on a joint footing. The roof of the National Stadium is covered by a double-layer membrane structure, with a transparent ETFE membrane fixed on the upper part of the roofing structure and a translucent PTFE membrane fixed on its lower part. A PTFE acoustic ceiling is attached to the side walls of the inner ring.

The National Stadium is a complex structure, posing great difficulties for its designers and constructors.

DESIGN CONCEPT

Architectural Summary

National Stadium is located on a gentle rise in the centre of the Olympic complex. An undulating composition of high and low elevations moderates the bulk of the vessel and gives it a dramatic sweeping form. The stadium's appearance is pure structure. Facade and structure are identical. The structural elements mutually support each other and converge into a grid-like formatio—almost like a bird's nest with its interwoven twigs. The design meets all the functional and technical requirements of an Olympic Stadium, but without communicating the insistent sameness of technocratic architecture dominated by large spans and digital screens. The spatial effect of the stadium is novel and radical and yet simple and of an almost archaic immediacy, thus creating a unique historical landmark for the Olympics 2008.

The plinth

The geometries of the plinth and stadium merge into one element, like a tree and its roots. Pedestrians flow on a lattice of smooth slate walkways that extend from the structure of the stadium. The spaces between walkways provide amenities for the stadium visitor: sunken gardens, stone squares, bamboo groves, mineral hills capes, and openings into the plinth itself. Gently, almost imperceptibly, the ground of the city rises and forms a plinth for the stadium. The entrance to the stadium is therefore slightly raised, providing a panorama of the entire Olympic complex.

The roof

Its appearance is pure structure. Facade and structure are identical. The structural elements mutually support each other and converge into a spatial grid – like formation, in which facades, stairs, bowl structure and the roof are integrated. To make the roof weather – proof the spaces in the structure of the stadium will be filled with a translucent membrane just as birds stuff their spaces between the woven twigs of their nests with soft filler. Since all of the facilities—restaurants, suites, shops and restrooms—are self – contained units, it is largely possible to do without a solid, enclosed facade. This allows natural ventilation of the stadium, which is the most important aspect of the stadium's sustainable design.

The bowl

The stadium is conceived as a large collective vessel, which makes a distinctive and unmistakable impression both when it is seen from a distance and from a nearby place. Inside the stadium, an evenly constructed bowl – like shape serves to generate crowd excitement and drive athletes to outstanding performances. To create a smooth and homogeneous appearance, the stands have minimal interruption and the acoustic ceiling hides the structure in order to focus attention on the spectators and the events on the field. The human crowd forms the architecture.

（3）建筑专业术语及词汇。

① 建筑面积 floor space

② 主体结构 main body

③ 设计使用年限 a design life of

④ 耐火等级 fire resistance capability

⑤ 抗震设防烈度 withstand an magnitude earthquake

⑥ 地下工程 underground project

⑦ 防水 water-resistance capability

⑧ 钢结构 steel structure

⑨ 空间马鞍形钢桁架编织式结构 saddle-shaped elliptic steel structure

⑩ 钢筋混凝土框架 concrete framework

⑪ 剪力墙结构体系 shear wall system

⑫ 双层膜结构 double-layer membrane structure

⑬ 透明的 ETFE 膜 transparent ETFE membrane

⑭ PTFE 声学吊顶 PTFE acoustic ceiling

⑮ 数码屏幕 digital screen

⑯ 体育场的空间效果 spatial effect of the stadium

⑰ 基座 plinth

⑱ 石板步道 slate walkway

⑲ 独自控制的单元 self-contained unit

⑳ 自然通风 natural ventilation

㉑ 公顷（等于2.471英亩）hectare

㉒ 临时的；暂时的；短暂的 temporary

㉓ 会场；（尤指）体育比赛场所 venue

㉔ 田径；田径赛 track and field

㉕ 设备 facility

㉖ 居民；（旅馆的）住宿者 resident

㉗ 巨大；重大；量级；（地震）级数 magnitude

㉘ 巨大的 colossal

㉙ （使）聚集 converge

㉚ 格子；（输电线路、天然气管道等的）系统网络 grid

㉛ 细枝；嫩枝 twig

㉜ 立场；基础；立足点 footing

㉝ 起伏的；呈波浪形的 undulating

㉞ 高地；海拔；提升 elevation

㉟（使）减轻；（使）缓和 moderate

㊱ 大部分；主要部分 the bulk of

㊲ 容器；船，飞船 vessel

㊳ 使（与……交织）interweave

㊴ 引人注意的；显眼的 insistent

㊵ 无感觉地；不知不觉地；极微地 imperceptibly

㊶ 立面 façade

㊷ 同一的，共同地 mutually

㊸ 融合 integrate

㊹ 可持续的 sustainable

㊺ 设想；想出 conceive

㊻ 集体的；共同的 collective

㊼ 最小的；最低限度的 minimal

㊽ 听觉的；音响的 acoustic

㊾ 观众 spectator

㊿ 平衡地 evenly

4.3.2.2　Text B 外国建筑——Eiffel Tower

埃菲尔铁塔

(1) 文本原文。

埃菲尔铁塔（法语：La Tour Eiffel）矗立在法国巴黎的战神广场，是世界著名建筑，法国文化象征与巴黎城市地标之一，也是巴黎最高建筑物，高 300 米，天线高 24 米，总高 324 米，于 1889 年建成，得名于设计它的著名建筑师、结构工程师古斯塔夫·埃菲尔。铁塔设计新颖独特，是世界建筑史上的技术杰作，是法国巴黎的重要景点和突出标志。

埃菲尔铁塔的结构体系既直观又简洁：底部是分布在每边 128 米

长底座上的 4 个巨型倾斜柱墩，倾角 54 度，由 57.63 米高度处的第一层平台连接支承。第一层平台和 115.73 米高度处的第二层平台之间是 4 个微曲的立柱，其向上转化为几乎垂直的、刚度很大的方尖塔，其间在 276.13 米高度处设有第三层平台。在 300.65 米高度处是塔顶平台，布置有电视天线。

铁塔总重 10000 吨，承担这些重量的是 4 个坚固的直伸至地下承力土层的沉箱基础。在 1884 年 6 月的时候，埃菲尔设计事务所的两位主要工程师埃米勒·努盖尔和毛里斯·科奇林就有了设计一座超高塔的思想。初步的设计像一个巨大的塔架，有四个由格构梁架构成的支腿，分立支撑在基础上并在塔顶收在一起，其间布置等间距横梁联系。为了更加符合公众的评价，努盖尔和科奇林邀请建筑师斯蒂芬·斯韦斯特雷对结构的表现形式作进一步处理。

斯韦斯特雷提议用石砌台座修饰塔脚，用具有纪念性的拱结构连接四个塔柱与第一层平台。每层平台设大型玻璃墙大厅，顶部采用球状造型等装饰手法来亮化结构。在其后的设计中工程师们进行了进一步的完善，像巨型拱之类的一些设计思想得到保留，并给予铁塔极富性格的表现。并且，塔柱优化以后，简化的横梁与线条相协调，这使铁塔结构更加简洁、优美、有力。

第一平台底部 4 个连接倾斜柱墩的大拱起初是为装饰用的，有人认为这 4 个拱破坏了塔结构的直线性、简洁性和"诚实"性，也损害了塔身的美观，但事实上这个"伪拱"已被公认为塔身外形的一个基本组成部分。如果说是横梁将四个塔柱简单联系在一起，那么，巨型拱的设置则把塔柱统一到铁塔这一整体中来。而且装饰拱的曲线线形与塔柱的曲线线形也很协调。从图上可以看出，部斜塔柱与水平横梁的装饰拱也具有一定的结构功能，虽然拱不承担塔柱向下传递的竖向荷载，但却使塔柱与横梁的连接加强，并使水平荷载作用下横梁与斜塔柱间的力流更加顺畅。从铁塔的初步设想图中可以看出，下横梁和塔柱的联结也表现出接头处曲线过渡的想法，只是在建筑师的建议下，通过不断的优化，使结构表现的更加充分，整体形式更加统一。

埃菲尔铁塔最初的建立是为了庆祝法国大革命胜利100周年，而后逐渐成为旅游景点，主要用于游客参观。同时埃菲铁塔也将在科学实验中起到巨大作用：空气动力学实验、材料耐力研究、登山生理学、无线电研究、电信问题、气象观测等等。120多年来埃菲尔铁塔已然经历了从技术到艺术进而转变为象征符号的这样一个过程，如今埃菲尔铁塔作为一种技术和艺术混合的工艺设计建筑，作为一种人文符号出现在世界大众眼前。

1880年法国刚刚摆脱普法战争中的耻辱，为了显示国力，1889年5月5日至11月6日，法国巴黎将再次举办世博会，主体是庆祝法国大革命胜利100周年。1886年5月，法国政府决定在巴黎战神广场设计一座高塔。条件有两个：高塔能吸引参观者买票参观；世博会后能轻易拆除。

1889年5月15日，为给世界博览会开幕典礼剪彩，铁塔的设计师古斯塔夫·埃菲尔亲手将法国国旗升上铁塔的300米高空，由此，人们为了纪念他对法国和巴黎的这一贡献，特别还在塔下为他塑造了一座半身铜像。

法国巴黎是浪漫之都，建筑物也都是低矮而且富有情调的，但是在市中心突然耸立起这个丑陋的、突兀的钢铁庞然大物，让巴黎市民很气愤，曾多次想拆除埃菲尔铁塔，认为它影响巴黎市容，是巴黎最糟糕、最失败的建筑物，而现在却成了法国甚至是全世界最吸金的建筑地标，2011年约有698万人参观，在2010年累计参观人数已超过2.5亿人次，每年为巴黎带来15亿欧元的旅游收入。巴黎人民也接受了它，并把埃菲尔铁塔作为法国的象征。法国人喜欢浪漫，他们不把这座庞然大物称作"大英雄"或"大丈夫"之类，而是将它亲密地爱称为"铁娘子"。埃菲尔铁塔经历了百年风雨，但在经过上世纪80年代初的大修之后风采依旧，巍然屹立在塞纳河畔。它是全体法国人民的骄傲，也是世界的骄傲。

埃菲尔铁塔是当时席卷世界的工业革命的象征。所以,它是为了世界博览会而落成的。庆祝法国革命胜利100周年,是代表法国荣誉的纪念碑。它成为当时席卷世界的工业革命的象征。它是世界建筑史上的技术杰作,曾经保持世界最高建筑记录45年。它成为法国和巴黎的一个重要景点,是现代巴黎的标志(巴黎圣母院可谓是古代巴黎的象征)。它显示出法国人异想天开式的浪漫情趣、艺术品位、创新魄力和幽默感。它代表着当时欧洲正处于古典主义传统向现代主义过渡与转换的特定时期。铁塔在第一次世界大战中在无线电通讯联络方面做出了重大贡献。

(2)参考译文。

Eiffel Tower (French: La Tour Eiffel) stands in the war god square in Paris, France. It is a world famous building, also one of the French cultural symbols, one of the Paris city landmarks, and the highest building in Paris. It is 300 meters high, the antenna is 24 meters high and the total height is 324 meters. It was built in 1889 and named from its designer, a famous architect and structural engineer, Gustav Eiffel. The design of the tower is novel and unique. It is a masterpiece of the world's architectural history and is an important scenic spot and prominent symbol of Paris, France.

The structure system of Eiffel Tower is intuitive and concise: the bottom is 4 huge tilted piers on each side of the 128 meters long base, the dip angle of which is 54 degree. They are connected and supported by the first floor platform of the 57.63m height. Between the first platform and the second platform at the height of the 115.73m there are 4 micro columns, the upper of which is curved slightly. The vertical and rigid obelisk is equipped with the third platform at the height of 276.13m, and at the height of 300.65m is a top platform with television antennas.

The gross weight of the tower is 10000t, and its load is carried by 4 caisson foundations which are strong and straight to the bottom bearing stratum. In June 1884, Emile Nouguier and Maurice Koechlin, the two leading engineers of Eiffel design office, had the idea of designing an ultra high tower. The preliminary design is like a huge tower, with four legs composed of a grid frame, supported on the basis of a vertical support and gathered together on the top of the tower. In order to be more agreeable to the public's evaluation, Nouguier and Koechlin invited architect Stephen Sauvestre to further deal with the manifestation of the structure.

Sauvestre proposes to use stone pedestal to trim the pagoda, with a commemorative arch structure connecting four tower columns and first floor platforms. Each platform has large glass wall hall, and the top uses decorative techniques such as spherical shape to brighten the structure. In the subsequent design, the engineers further perfected it: some of the design ideas such as giant arches were preserved, to show the very character of the tower. In addition, the setting of the beam is simplified and coordinated with the line shape after the column optimization, which makes the structure of the tower more concise, more graceful and more powerful.

4 large arches at the bottom of the first platform are used for decoration at first. Some people think that these 4 arches destroy the straightness, simplicity and honesty of the tower structure, and damage the beauty of the tower as well. In fact, this "pseudo arch" has been recognized as a basic component of the shape of the tower. If the beams are simply linked together by four columns, the installation of the giant arch will bring the tower column into the whole of the tower. Moreover, the curve line of the decorative arch is also in line with the curve alignment of the tower column. As can be seen from the graph, the decorative arch of the column and the horizontal beam also has a certain structural function. Although the arch does not bear the vertical load of the column downward, the connection of the column and the beam is strengthened, and the force flow between the horizontal

beam and the tower column is more smooth. From the preliminary picture of the tower, it can be seen that the connection of the lower beam and the tower column also shows the idea of the transition of the curve at the joint. It is only under the suggestion of the architect to make the structure more fully, and the whole form more unified through continuous optimization.

Eiffel Tower was originally built to celebrate the victory of the French Revolution 100th anniversary, and then gradually has become a tourist attraction, mainly for tourists. At the same time, Eiffel tower will also play a great role in scientific experiments: Aerodynamics experiments, material endurance research, mountaineering physiology, radio research, telecommunications, meteorological observation and so on. For more than 120 years, Eiffel Tower has experienced a process from technology to art to symbolism. Today, Eiffel Tower is a technology and art mixed process design architecture. As a humanistic symbol, it appears in the eyes of the world.

In 1880, France just got rid of the humiliation in the Prussian War. In order to show the national strength, from May 5th, 1889 to November 6th, Paris, France, would hold World Expo again, the main part of which was to celebrate the victory of the French Revolution 100th anniversary. In May 1886, the French government decided to design a tower in Ares square in Paris. There were two conditions: the tower could attract visitors to buy tickets; after the Expo, it could be easily demolished.

on May 15, 1889, in order to cut the opening ceremony of the World Expo, Gustav Eiffel, the designer of the tower, raised the French flag at the height of 300 meters in the tower. In order to commemorate his contribution to France and Paris, people in particular also created a half bronze statue for him under the tower.

Because Paris is a romantic capital in French, the buildings are low and full of sentiment, while the ugly, abrupt iron and steel giant that suddenly rises in the center of the city has made Paris people angry. It has repeatedly tried to dismantle the Eiffel Tower, considering it is in the Paris City, the worst and most failed construction of Paris. It has now become the most gold-sucking landmark in France, even in the world. In 2011, about 6980000 people visited more than two hundred and fifty million people in 2010, bringing 1500000000 in tourism revenue to Paris each year. The people of Paris accepted it and regarded Eiffel Tower as a symbol of France. French people like romance. They do not call this giant a "big hero" or "big husband". Instead, they call it "Iron Lady" intimately. Eiffel Tower has gone through a hundred years of wind and rain, but after the overhaul of the early 80s of the last century, it still stood on the Seine River (法语: La seine). It is the pride of all the French people and the pride of the world.

Eiffel Tower was the symbol of the industrial revolution that swept the world at that time. So it was built for the world exposition. Celebrating the victory of the French Revolution 100th anniversary is a monument to France's honor. It became the symbol of the industrial revolution that swept the world at that time. It is a technical masterpiece in the history of World Architecture, and it has been the tallest building in the world for 45 years. It has become an important scenic spot in France and Paris, and is a symbol of modern Paris (Notre Dame de Paris is a symbol of ancient Paris). It shows the French's fantasies of romantic taste, artistic taste, creativity and sense of humor. It represents the specific period when Europe was in the transition and transformation from classicalism to modernism. The tower has made significant contributions to radio communications in the First World War.

(3) Terms and Vocabularies。

① cultural symbol 文化象征

② landmark 地标

③ architect 建筑师

④ structural engineer 结构工程师

⑤ structure system 结构体系

⑥ base 底座

⑦ dip angle 倾角

⑧ obelisk 方尖塔

⑨ gross weight 总重

⑩ caisson 沉箱

⑪ stratum 地层

⑫ preliminary design 初步的设计

⑬ grid frame 格构梁架

⑭ manifestation of the structure 结构的表现形式

⑮ stone pedestal 石砌台座

⑯ pagoda 宝塔

⑰ arch structure 拱结构

⑱ glass wall hall 玻璃墙大厅

⑲ spherical shape 球状造型

⑳ beam 横梁

㉑ masterpiece 杰作；名作

㉒ prominent 杰出的；著名的

㉓ intuitive 直觉的；直观的

㉔ concise 简明的；简洁的

㉕ tilted 倾斜的；翘起的

㉖ pier 码头；防波堤；桥墩；窗间壁

㉗ vertical 垂直的；竖立的；垂直线，垂直面；[建] 竖杆；垂直位置

㉘ bearing 支座

㉙ agreeable 令人愉快的；惬意的；适合的，一致的

㉚ evaluation 评价；评估

㉛ manifestation 表示；显示

㉜ spherical 球形的；球面的；天体的；天空的

㉝ optimization 最佳化；最优化；优选法；优化组合

㉞ installation 安装；装置

㉟ transition 过渡；转变；变迁

㊱ curve 弧线；曲线；弯曲物

㊲ joint 接合处

㊳ physiology 生理学；生理机能

㊴ symbolism 象征主义；象征手法

㊵ demolish 摧毁；推翻；拆毁（尤指大建筑物）

㊶ commemorate 纪念，庆祝；成为…的纪念

㊷ sentiment 感情，情绪；情操；意见，观点；感伤

㊸ dismantle 拆开；拆卸；废除；取消

㊹ intimately 熟悉地；亲密地；私下地；谙熟地

㊺ overhaul 检查；彻底检修；详细检查；大修

㊻ exposition 博览会；展览会

㊼ anniversary 周年纪念日

㊽ monument 纪念碑；遗迹；遗址；丰碑

㊾ classicalism 古典主义

㊿ modernism 现代主义

4.3.3 翻译强化训练

4.3.3.1 句子翻译

1. 汉译英

① 国家体育场位于北京奥林匹克公园中心区南部，为2008年第

29 届奥林匹克运动会的主体育场。

② 奥运会、残奥会开闭幕式、田径比赛及足球比赛决赛在这里举行。

③ 国家体育场工程主体结构设计使用年限 100 年，耐火等级为 1 级，抗震设防烈度 8 度，地下工程防水等级 1 级。

④ 钢结构与混凝土看台上部完全分开，互不相连。形式上呈相互围合，基础则坐在一个相连的基础底板上。

⑤ 国家体育场工程结构复杂，难度极高，对设计者和建设者来说是一个不小的挑战。

⑥ 国家体育场坐落于奥林匹克公园建筑群的中央位置，地势略微隆起，如同巨大的容器。

⑦ 高低起伏波动的基座缓和了容器的体量，而且给了它戏剧化的弧形外观。

⑧ 各个结构元素之间相互支撑，汇聚成网格状——就如同一个由树枝编织成的鸟巢。

⑨ 在满足奥运会体育场所有的功能和技术要求的同时，设计上并没有被那些类似的过于强调建筑技术化的大跨度结构和数码屏幕所主宰。

⑩ 体育场的空间效果新颖前卫，但又简洁古朴，从而为 2008 年奥运会创造了独史无前例的地标性建筑。

⑪ 基座与体育场的几何体合二为一，如同树根与树。

⑫ 行人走在平缓的格网状石板步道上，步道延续了体育场的结构肌理。

⑬ 体育场的入口处地面略微升高，因此，可以浏览到整个奥林匹克公园建筑群的全景。

⑭ 各个结构元素之间相互支撑，汇聚成网格状，就像编织一样，将建筑物的立面、楼梯、碗状看台和屋顶融合为一个整体。

⑮ 如同鸟会在它们树枝编织的鸟巢间加一些软充填物一般，为

了使屋顶防水，体育场结构间的空隙将被透光的膜填充。

⑯ 由于所有的设施——餐厅，客房，商店和卫生间都是独自控制的单元，所以建筑外立面的整体封闭因此是非常不必要的。

⑰ 这使体育场可以做到自然通风，是体育场环保设计的最重要的一个方面。

⑱ 体育场被设计成为巨大的人群的容器，无论远眺还是近观，都给人留下与众不同的、不可磨灭的印象。

⑲ 体育场内部，这种均匀的碗状结构形体将能调动观众的兴奋情绪，有可能使运动员超水平发挥。

⑳ 创造连贯一致的外表，座位的干扰被控制到最小，声学吊顶将结构遮掩使得观众和场地上的活动成为注意焦点。

2. 英译汉

① It is a masterpiece of the world's architectural history and is an important scenic spot and prominent symbol of Paris, France.

② Between the first platform and the second platform at the height of the 115.73m there are 4 micro columns, the upper of which is curved slightly. The vertical and rigid obelisk is equipped with the third platform at the height of 276.13m, and at the height of 300.65m is a top platform with television antennas.

③ The preliminary design is like a huge tower, with four legs composed of a grid frame, supported on the basis of a vertical support and gathered together on the top of the tower.

④ Each platform has large glass wall hall, and the top uses decorative techniques such as spherical shape to brighten the structure.

⑤ In addition, the setting of the beam is simplified and coordinated with the line shape after the column optimization, which makes the structure of the tower more concise, more graceful and more powerful.

⑥ Although the arch does not bear the vertical load of the column downward, the connection of the column and the beam is strengthened, and the force flow between the horizontal beam and the tower column is smoother.

⑦ From the preliminary picture of the tower, it can be seen that the connection of the lower beam and the tower column also shows the idea of the transition of the curve at the joint. It is only under the suggestion of the architect to make the structure more fully, and the whole form more unified through continuous optimization.

⑧ Eiffel Tower was originally built to celebrate the victory of the French Revolution 100th anniversary, and then gradually has become a tourist attraction, mainly for tourists.

⑨ For more than 120 years, Eiffel Tower has experienced a process from technology to art to symbolism. Today, Eiffel Tower is a technology and art mixed process design architecture. As a humanistic symbol, it appears in the eyes of the world.

⑩ In May 1886, the French government decided to design a tower in Ares square in Paris. There were two conditions: the tower could attract visitors to buy tickets; after the Expo, it could be easily demolished.

⑪ In May 15, 1889, in order to cut the opening ceremony of the World Expo, Gustav Eiffel, the designer of the tower, raised the French flag at the height of 300 meters in the tower. In order to commemorate his contribution to France and Paris, people in particular also created a half bronze statue for him under the tower.

⑫ Because Paris is a romantic capital in French, the buildings are low and full of sentiment, while the ugly, abrupt iron and steel giant that suddenly rises in the center of the city has made Paris people angry.

⑬ It has now become the most gold-sucking landmark in France, even in the world. In 2011, about 6980000 people visited more than two hundred and fifty million people in 2010, bringing 1500000000 in tourism revenue to Paris each year.

⑭ The people of Paris accepted it and regarded Eiffel Tower as a symbol of France, and it has now become the most gold – sucking landmark in France, even in the world.

⑮ Eiffel Tower has gone through a hundred years of wind and rain, but after the overhaul of the early 80s of the last century, it still stood on the Seine River（法语：La seine）. It is the pride of all the French people and the pride of the world.

⑯ It became the symbol of the industrial revolution that swept the world at that time.

⑰ Eiffel Tower is a technical masterpiece in the history of World Architecture, and it has been the tallest building in the world for 45 years.

⑱ It shows the French's fantasies of romantic taste, artistic taste, creativity and sense of humor.

⑲ It represents the specific period when Europe was in the transition and transformation from classicalism to modernism.

⑳ The tower has made significant contributions to radio communications in the First World War.

4.3.3.2 段落翻译

1. 汉译英

① 中国传统建筑与传统西方建筑最显着的区别在于建筑材料。大多数古老的西方建筑都是用石头建造的，庄严而富丽堂皇。最重要的是，他们今天得以幸存下来。中国古代人非常善于使用木材，创造了复杂的榫卯结构，为中国古代木结构带来独特的美学。中国古代的大多数宫殿，寺庙和宝塔都是用木头建造的。它们节能、环保且耐用，但在历史上很容易被火烧毁。幸运的是，今天仍有许多中国古代

木制建筑宝石被遗留下来。

② 四合院是一座建筑群,由周围的四栋房屋组成。四合院是中国古代建筑的典型形式,尤其是在中国北方。而北京四合院是最经典的。四合院是封闭的,朝内的,以保护他们免受寒冷的冬季风和春天的沙尘暴。四合院按行排列,根据居民的社会地位而有所不同。高级官员和富商的大四合院都是专门建造的,屋顶横梁和柱子雕刻精美,每个都有前院和后院。通常情况下,大门内有一道屏风墙,以便外人无法直接看到庭院,并保护房屋免受恶灵的侵害。在一些大的四合院门外,常见的是找到一对石狮子。大门通常涂有朱红色,并有大型铜门环。

③ 因为南部、北部、西部和东部的自然环境完全不同,建筑材料也不同。根据自身的生产需求和生活需求,他们采用自然环境提供的材料来构建不同的建筑,并创造自己的建筑风格。在中国北方省份的黄土覆盖地区,古代人以黄土为材料建房,然后用黄土制砖,建筑用黄土制成的砖块方便耐用。在中国的南部,天气潮湿,人们用竹子建造建筑物。在蒙古地区,根据游牧民族的习惯,蒙古包是他们特有的。蒙古包很容易拆卸和移动。总而言之,古代人们希望利用大自然所提供的材料来建造不同风格的建筑,为中国古代建筑做出贡献。

2. 英译汉

① Neoclassical architecture is a kind of architectural style, which originated from the Neoclassical Movement in the mid – 18th century. As a reaction to the anti – structural decoration of the Rococo style and byproduct of some imitative classical features in the later Baroque, it is characterized by regular composition and the pursuit of grandeur and rigor. The neoclassical style is an attempt to revive the Classical architecture with a wish to return to the Ancient Greece, Ancient Rome and to extend the Renaissance architecture. The movement lasted around a century (1750s – 1850s).

② The interior of the Parthenon is surrounded by by large niches which housed statues of Roman Gods. It is round so as to place all gods at the same level of importance. The seven magnificent niches are situated between 2 Corinthian columns where every niche held a statue. These statues were believed to represent the seven gods linked to worship of the planets. The niche in front of the entrance stands out from the rest due to it's greater size and different framing. The seven astrological planets known to the ancient Romans were the Sun, Moon, Mars, Mercury, Jupiter, Venus, and Saturn. It therefore would be a reasonable assumption that the Roman gods affiliated to these planets were depicted in the Pantheon statues.

③ Every person who has studied architecture, whether he or she is an architect or not in his or her later profession, must know the name of Le Corbusier. Corbusier is among the pioneers and representatives of modernist architecture. By subverting the classical forms that have been inherited for thousands of years, the masters of his generation, coming from the classics, create brand – new architectural types corresponding to the development of the industrial era and have been influencing today. What we see today is modern architecture of different styles from classical architecture, which is benefited from the sublation and innovation of the masters of that generation.

4.3.4 练习答案

4.3.4.1 句子翻译

1. 汉译英答案

① Located at the southern part of the Olympic Park in Beijing, the National Stadium is the main stadium of the 29th Olympic Games in 2008.

② The venue hosted the opening and closing ceremonies of the Beijing

Olympic Games and Paralympic Games, the track and field competitions, and the football finals.

③ The main body of the National Stadium has a design life of 100 years. Its fire resistance capability is first – rate, and it can withstand an eight – magnitude earthquake. The water – resistance capability of its underground project is also first – rate.

④ Seemingly mutually enclosed, the upper part of the concrete stand and the stadium's steel structure are separated from each other, but both are based on a joint footing.

⑤ The National Stadium is a complex structure, posing great difficulties and challeges for its designers and constructors.

⑥ National Stadium is located on a gentle rise in the centre of the Olympic complex like a huge vessel.

⑦ An undulating composition of high and low elevations moderates the bulk of the vessel and gives it a dramatic sweeping form.

⑧ The structural elements mutually support each other and converge into a grid – like formation—almost like a bird's nest with its interwoven twigs.

⑨ The design meets all the functional and technical requirements of an Olympic Stadium, but without communicating the insistent sameness of technocratic architecture dominated by large spans and digital screens.

⑩ The spatial effect of the stadium is novel and radical and yet simple and of an almost archaic immediacy, thus creating a unique historical landmark for the Olympics 2008.

⑪ The geometries of the plinth and stadium merge into one element, like a tree and its roots.

⑫ Pedestrians flow on a lattice of smooth slate walkways that extend from the structure of the stadium.

⑬ The entrance to the stadium is therefore slightly raised, providing a panorama of the entire Olympic complex.

⑭ The structural elements mutually support each other and converge into a spatial grid – like formation, in which facades, stairs, bowl structure and the roof are integrated.

⑮ To make the roof weatherproof the spaces in the structure of the stadium will be filled with a translucent membrane just as birds stuff their spaces between the woven twigs of their nests with soft filler.

⑯ Since all of the facilities—restaurants, suites, shops and restrooms—are self – contained units, it is largely possible to do without a solid, enclosed facade.

⑰ This allows natural ventilation of the stadium, which is the most important aspect of the stadium's sustainable design.

⑱ The stadium is conceived as a large collective vessel, which makes a distinctive and unmistakable impression both when it is seen from a distance and from a nearby place.

⑲ Inside the stadium, an evenly constructed bowl – like shape serves to generate crowd excitement and drive athletes to outstanding performances.

⑳ To create a smooth and homogeneous appearance, the stands have minimal interruption and the acoustic ceiling hides the structure in order to focus attention on the spectators and the events on the field.

2. 英译汉答案

① 铁塔设计新颖独特,是世界建筑史上的技术杰作,是法国巴黎的重要景点和突出标志。

② 第一层平台和115.73米高度处的第二层平台之间是4个微曲的立柱,其向上转化为几乎垂直的、刚度很大的方尖塔,其间在276.13米高度处设有第三层平台。在300.65米高度处是塔顶平台,布置有电视天线。

③ 初步的设计像一个巨大的塔架，有四个由格构梁架构成的支腿，分立支撑在基础上并在塔顶收在一起，其间布置等间距横梁联系。

④ 每层平台设大型玻璃墙大厅，顶部采用球状造型等装饰手法来亮化结构。

⑤ 并且，塔柱优化以后，简化的横梁与线条相协调，这使铁塔结构更加简洁、优美、有力。

⑥ 虽然拱不承担塔柱向下传递的竖向荷载，但却使塔柱与横梁的连接加强，并使水平荷载作用下横梁与斜塔柱间的力流更加顺畅。

⑦ 从铁塔的初步设想图中可以看出，下横梁和塔柱的联结也表现出接头处曲线过渡的想法，只是在建筑师的建议下，通过不断的优化，使结构表现的更加充分，整体形式更加统一。

⑧ 埃菲尔铁塔最初的建立是为了庆祝法国大革命胜利 100 周年，而后逐渐成为了旅游景点，主要用于游客参观。

⑨ 120 多年来埃菲尔铁塔已然经历了从技术到艺术进而转变为象征符号的这样一个过程，如今埃菲尔铁塔作为一种技术和艺术混合的工艺设计建筑，作为一种人文符号出现在世界大众眼前。

⑩ 1886 年 5 月，法国政府决定在巴黎战神广场设计一座高塔。条件有两个：高塔能吸引参观者买票参观；世博会后能轻易拆除。

⑪ 1889 年 5 月 15 日，为给世界博览会开幕典礼剪彩，铁塔的设计师古斯塔夫·埃菲尔亲手将法国国旗升上铁塔的 300 米高空，由此，人们为了纪念他对法国和巴黎的这一贡献，特别还在塔下为他塑造了一座半身铜像。

⑫ 法国巴黎是浪漫之都，建筑物也都是低矮而且富有情调的，但是在市中心突然耸立起这个丑陋的、突兀的钢铁庞然大物，让巴黎市民很气愤。

⑬ 现在却成了法国甚至是全世界最吸金的建筑地标，2011 年约有 698 万人参观，在 2010 年累计参观人数已超过 2.5 亿人次，每年

为巴黎带来15亿欧元的旅游收入。

⑭ 巴黎人民也接受了它,并把埃菲尔铁塔作为法国的象征,成为法国甚至是全世界最吸金的建筑地标。

⑮ 埃菲尔铁塔经历了百年风雨,但在经过上世纪80年代初的大修之后风采依旧,巍然屹立在塞纳河畔。它是全体法国人民的骄傲,也是世界的骄傲。

⑯ 埃菲尔铁塔是当时席卷世界的工业革命的象征。

⑰ 它是世界建筑史上的技术杰作,曾经保持世界最高建筑记录45年。

⑱ 它显示出法国人异想天开式的浪漫情趣、艺术品位、创新魄力和幽默感。

⑲ 它代表着当时欧洲正处于古典主义传统向现代主义过渡与转换的特定时期。

⑳ 铁塔在第一次世界大战中在无线电通讯联络方面做出了重大贡献。

4.3.4.2 段落翻译

1. 汉译英答案

① Most significant difference between traditional Chinese architecture and traditional western architecture is the construction material. Most ancient Western buildings were built up with stones, solemn and magnificent. Most importantly, they survive today. Ancient Chinese people were very good at using wood, and created the complex mortise and tenon joint structure to bring in the unique aesthetics for ancient Chinese wooden architecture. Most of the palaces, temples and pagodas of ancient China were built with wood. They were energy-efficient, environmentally-friendly and durable, but were easily destroyed by fire in history. Fortunately, there are still many ancient Chinese wooden architecture gems left and preserved today.

② The Siheyuan is a building complex formed by four houses around a quadrangular courtyard. The siheyuan is a typical form of ancient Chinese architecture, especially in the north of China. And Beijing Siheyuan is the most classic. The Siheyuan are enclosed and inward facing to protect them from the harsh winter winds and the dust storms of spring. Their design also reflects the traditions of China, following the rules of Feng Shui and the patriarchal, Confucian tenants of order and hierarchy that were so important to society. Siheyuans are arranged in rows and vary in size and design according to the social status of the residents. The big Siheyuans of high-ranking officials and wealthy merchants were specially built with roof beams and pillars are beautifully carved and painted, each with a front yard and back yard. Normally, there is a screen-wall inside the gate so that outsiders cannot see directly into the courtyard and to protect the house from evil spirits. Outside the gate of some large Siheyuan, it is common to find a pair of stone lions. The gates are usually painted vermilion and have large copper door rings.

③ In China, just as the natural environment is totally different, so are the building materials. According to their own production needs and living needs, they adopt the materials the natural environment provides to build different architecture, and create their own architecture style. In loess covered areas of China's Northern provinces, ancient people use loess to make brick, and the building using bricks made of loess is convenient and durable. In the southern part of China, the weather is wet and humid, people build building by bamboo. In Mongolian area, according to the habit of nomadic people, yurt is specific to them. The yurt is easy to disassemble and move. All in all, ancient people would like to use the materials the nature provides to build, which make the architectures with different styles, and make their contributions to Chinese ancient architecture.

2. 英译汉答案

① 新古典主义建筑是一种建筑风格，起源于18世纪中叶的新古典主义运动。作为对洛可可风格反构造装饰的反应，以及巴洛克晚期中一些仿古典特征的副产物，其特点是构图规整，追求雄伟、严谨。新古典主义风格尝试复兴古典建筑，希望回到古希腊、古罗马，同时扩展文艺复兴时期的建筑。该运动持续了大约一个世纪（1750~1850）。

② 帕特农神庙的内部被巨大的壁龛所包围，里面有罗马神的雕像。它呈圆形，为的是把所有的神放在同等重要的水平上。这七个宏伟的壁龛位于两个科林斯柱之间，每一个小壁龛都有一座雕像。人们认为这些雕像代表了与行星崇拜有关的七位神。入口前的壁龛因其更大的尺寸和不同的框架而令人瞩目。古罗马人所知的七颗星象行星是太阳、月亮、火星、水星、木星、金星和土星。因此，可以合理设想，万神殿雕像即是描绘的附属于这些行星的罗马诸神。

③ 每个读过建筑学的人，无论他/她后来从事的职业是否是建筑师，一定都会知晓柯布西耶的大名。柯布西耶是现代主义建筑的先锋和代表人物之一。他们那一代大师，从古典中走来，却能颠覆千百年沿袭下来的古典形式，创造出对应工业时代发展的崭新的建筑类型并一直影响至今。今天我们所看到的与古典建筑形式迥异的各种风格的现代建筑，正是得益于那一代大师们的扬弃和创新。

4.4　翻译技巧：定语从句的译法

建筑文本：中国建筑——天坛

外国建筑——美国帝国大厦

4.4.1　翻译技巧——定语从句

英语定语从句的译法主要涉及到限制性定语从句的译法和非限制

性定语从句的译法。此外，有些英语定语从句和主句之间还存在着状语关系，对这种定语从句的译法亦值得探讨。

4.4.1.1 限制性定语从句的译法

限制性定语从句对所修饰的先行词起限制作用，与先行词关系密切，不用逗号分开。翻译这类句子时往往可以用。

1. 前置法

把英语限制性定语从句译成带"的"的定语词组，放在被修饰词之前，从而将复合句译成汉语单句。

① I was, to borrow from John Le Carre, the spy who was to stay out in the cold.

借用约翰·勒卡雷的话来说，我成了一个被打入冷宫的间谍了。

② It is a consolation to know that they will surely carry on the cause for which Edgar Snow strove so faithfully all his life.

了解到他们一定会把斯诺终身为之奋斗不渝的事业继承下来，这是一件令人快慰的事情。

2. 后置法

上述译成前置定语的方法大都适用于限制性定语从句，但一般用于翻译比较简单的英语定语从句，如果从句结构复杂，译成汉语前置定语显得太长而不符合汉语表达习惯时，往往可以译成后置的并列分句。

（1）译成并列分句，重复英语先行词。

① He unselfishly contributed his uncommon talents and indefatigable spirit to the struggle which today brings them (those aims) within the reach of a majority of the human race.

他把自己非凡的才智和不屈不挠的精神无私地献给了那次斗争，那次斗争今天已使人类大多数人实现了那些目标。

② They are striving for the ideal which is close to the heart of every Chinese and for which, in the past, many Chinese have laid down their lives.

他们正在为实现一个理想而努力,这个理想是每个中国人所珍爱的,在过去,许多中国人曾为了这个理想而牺牲了自己的生命。

(2) 译成并列分句,省略英语先行词。

① A good deal went on in the steppe which he—her father—did not know.

草原上发生了许多事情,是他和她的父亲都并不知道的。

② He managed to raise a crop of 200 miracle tomatoes that weighed up to two pounds each.

他居然种出了200个奇迹般的西红柿,每个重达2磅。

3. 溶合法

(1) 溶合法是指把原句中的主语和定语从句溶合在一起译成一个独立句子的一种翻译方法。由于限制性定语从句与主句关系紧密,所以溶合法比较适用于翻译限制性定语从句。英语中的 There be…结构汉译时往往就是这样处理的。

① There are many people who want to see the film.

许多人要看这部电影。

② There was another man who seemed to have answers, and that was Robert McNamara.

另外一个人似乎胸有成竹,那就是是麦克纳马拉。

(2) 此外,还有些带定语从句的英语复合句,译成汉语时可将英语主句压缩成汉语词组做主语,而把定语从句译成谓语,溶合成一个句子。

① "We are a nation that must beg to stay alive," said a foreign economist.

一位外国经济学家说道,"我们这个国家不讨饭就活不下去"。

② We used a plane of which almost every part carried some indication of national identity.

我们驾驶的飞机几乎每一个部件都有国籍的某些标志。

例句——汉译英

① 中国是古代文化和现代文化的摇篮,这种文化以理想鼓舞着全世界。

China is the cradle of an ancient and modern culture that has inspired the world with ideals.

② 许多建筑都带有明显的地域文化特点。

There are many architectural styles which are apparently characterized with their geography and culture.

③ 从东门进入,经风门厅、过厅到达中央大厅。中心大厅占地3600平方米,护墙和地面用彩色大理石铺砌。

Entering through the East Gate, and passing through two halls, one will reach the Central Hall which has a colorful marble floor and walls, covering an area of 3,600 square meters.

④ 蓝室的阳台面对南院草坪,是总统与贵宾经常向民众挥手露面的地方。

The balcony of the blue room faces the lawn of the South Lawn. It is the place where the president and VIP often waved to the public.

⑤ 林肯卧室是林肯办公和召开内阁会议的地方,著名的《解放黑人奴隶宣言》即在此签字。

Lincoln bedroom is the place where Lincoln works and holds cabinet meetings. The Emancipation proclamation was signed here.

⑥ 正面墙上是身着戎装威容凛然的华盛顿油画像,两边摆着两只雅致的中国古瓷花瓶。办公室左边墙架上陈设的外国贵宾赠送的礼物中,有中国1979年赠送仿古青铜器。

On the front wall is a Washington painting with an impressive military dress and two elegant Chinese antique vases on both sides. Among the gifts presented by foreign dignitaries on the left wall of the office are the antique bronzes which were donated by China in 1979.

例句——英译汉

① Chinese architecture constitutes the only system based mainly on wooden structures of unique charming appearance. This differs from all other architectural systems in the world which are based mainly on bricks and stone structures.

中国建筑是以独具外形魅力的木质结构构成的单一系统,这有别于世界上其他砖石结构为主的建筑系统。

② As the main representative of the British medieval architecture, the architectural style and characteristics of the Westminster Abbey were constantly changing during the marathon era, from the Norman, Gothic, to the early Renaissance style, but its basic features still belonged to Gothic style, so it took over 700 years of repair. And it was thanks to an architect like Scott who kept his original appearance.

作为英国中世纪建筑的主要代表,威斯敏斯特教堂的建筑风格和特点,虽然在马拉松式的建造年代中不断地推移变化,从诺曼式、哥特式,一直到早期文艺复兴的式样,不过它的基本特色仍属于哥特式,所以历经700多年的修葺犹能保持原貌,实在多亏了斯科特这样的建筑师。

③ The cathedral architecture is Gothic, with several colored glass inlaid spire tied together. There is also a crowded cemetery in the middle of the cathedral where there are many great figures buried.

教堂建筑为哥特式,数个由彩色玻璃嵌饰的尖顶并列在一起,显得别致动人。教堂中间还有一处拥挤的墓场,埋葬有诸多伟大人物。

④ "Walking into Westminster Abbey is not a monarch's mausoleum, but a monument to the nation to thank the greatest people who have added to the country," he said. "This is the respect of the British people for their talents."

他曾感慨道:"走进威斯敏斯特教堂,人们所瞻仰的不是君王们的陵寝,而是国家为感谢那些为国增光的最伟大人物的纪念碑。这便是英国人民对于有才华的人的尊敬。"

⑤ The Westminster Abbey is the place where the kings are crowned, the wedding ceremony is held, and the Royal Mausoleum of the UK is the place where the Westminster is a stone history book of the British royal family.

威斯敏斯特教堂是历代国王加冕登基、举行婚礼庆典的地方,也是英国的王室陵墓所在地,可以说西敏寺是一部英国王室的石头史书。

4.4.1.2 非限制性定语从句的译法

英语非限制性定语从句对先行词不起限制作用,只对它加以描写、叙述或解释。翻译这类从句时可以运用。

1. 前置法

一些较短而具有描写性的英语非限制性的定语从句,也可译成带"的"的前置定语,放在被修饰词前面,但这种处理方法不如用在英语限制性定语从句那样普遍。

① The sun, which had hidden all day, now came out in all its splendor.

那个整天躲在云层里的太阳,现在又光芒四射地露面了。

② The emphasis was helped by the speaker's mouth, which was wide, thin and hard set.

讲话人那又阔又薄又紧绷绷的嘴巴,帮助他加强了语气。

2. 后置法

(1) 译成并列分句。

在译文中把原文从句后置,重复英语关系词所代表的含义。在译文中从句后置,省略英语关系词所代表的含义。

① I told the story to John, who (= and he) told it to his brother.

他把这件事告诉了约翰,约翰又告诉了他的弟弟。

② He saw in front that haggard white-haired old man, whose eyes flashed red with fury.

他看见前面那个憔悴的白发老人,眼睛里愤怒地闪着红光。

(2) 译成独立句。

① He had talked to Vice-President Nixon, who assured him that everything that could be done would be done.

他和副总统尼克松谈过话。副总统向他担保,凡是能够做到的都将竭尽全力去做。

② One was a violent thunderstorm, the worst I had ever seen, which obscured my objective.

有一次是暴风骤雨,猛烈的程度实为我生平所仅见。这阵暴风雨遮住了我的目标。

(3) 兼有状语职能的定语从句。

英语中有些定语从句,兼有状语从句的职能,在意义上与主句有状语关系,说明原因、结果、目的、让步、假设等关系。翻译时应善于从原文的字里行间发现这些逻辑上的关系,然后译成汉语各种相应的偏正复句。

① The ambassador was giving a dinner for a few people whom he wished especially to talk to or to hear from.

大使只宴请了几个人,因为他特地想和这些人谈谈,听听他们的意见。

② There was something original, independent, and heroic about the plan that pleased all of them.

这个方案富于创造性,独出心裁,很有魄力,所以使他们都很喜欢。

③ Men become desperate for work, any work, which will help them to keep alive their families.

人们极其迫切地要求工作，不管什么工作，只要它能维持一家人的生活就行。

例句——汉译英

① 国家游泳中心赛后将成为北京最大的水上乐园，所以设计者针对各个年龄层次的人，探寻水可以提供的各种娱乐方式，开发出水的各种不同的用途，他们将这种设计理念称作"水立方"。

The National Swimming Center will be the largest water park in Beijing after the race, so the designer seeks the various forms of entertainment that water can provide for people of all ages and develop various uses of the water, which they call the "Water Cube".

② 第一，中国古代建筑十分注重整体布局与左右对称，象征了秩序与稳定。

Ancient Chinese architecture puts particular emphasis on the general layout and bilateral symmetry, which suggest order and stability.

③ 据说，最早的苏州园林起源于公元前6世纪春秋时期吴国国王修建的皇家园林。

It is said that the earliest Suzhou gardens date from the Spring and Autumn Period in the 6th century BC, when one king of Wu built his royal garden.

④ 体育场被设计成为巨大的人群的容器，无论远眺还是近观，都给人留下与众不同的、不可磨灭的印象。

The stadium is conceived as a large collective vessel, which makes a distinctive and unmistakable impression both when it is seen from a distance and from a nearby place.

例句——英译汉

① The establishment of the church seems to follow the advice of St.

Peter, who is said to have appeared at the canon of the first bishop of **MERIS**. The process of Norman conqueror William's invasion of England in Eleventh Century was vividly recorded in detail from the bayai color map (coloured textile and landscape murals).

这座教堂的建立似乎是遵循圣彼得的指示，据说他曾在首位主教梅里图斯的封圣典礼上现身。巴耶彩图（彩织情景壁画）生动详细地记录了11世纪诺曼征服者威廉侵占英格兰的过程。

② A great number of ecclesiastical buildings remain from this period, of which many of the larger churches are considered priceless works of art and are inscribed on the UNESCO World Heritage List.

这一时期众多的基督教会建筑保留至今，其中很多大型教堂被视为珍贵的艺术品，被联合国教科文组织列入世界遗产名录。

③ The Colosseum has four stories and each of the first three stories has 80 arched entrances, which enabled spectators to enter and exit in good order.

罗马竞技场有四层，下面三层每层都有80个拱形入口，可供观众井然有序地进出竞技场。

④ The complex ranks among the most remarkable Romanesque structures from medieval Europe, which was put on the World Heritage List in 1987.

这组建筑是中世纪欧洲杰出的罗马式建筑典范，1987年被列入世界遗产名录。

⑤ Galileo, the great Italian physicist and astronomer, is said to have conducted his revolutionary experiments with falling objects from the leaning tower, which undoubtedly adds more charm to the Tower.

据说，意大利伟大的物理学家、天文学家伽利略曾在斜塔上做过具有重大意义的自由落体实验，这无疑给比萨斜塔增添了些许魅力。

4.4.2 建筑文本翻译实训

4.4.2.1 文本 A：中国建筑——天坛

(1) 文本原文。

天坛

天坛是一个值得参观的地方。天坛位于北京紫禁城的东南方,是明、清两朝皇帝举行祭天大典的场所,占地 273 公顷,平面布局北圆南方,象征天圆地方。两道坛墙将坛域分为内、外坛,祭祀建筑集中于内坛,并分为南北两部分。南部为"圜丘坛",北部为"祈谷坛"。南北两坛有一条长 360 米的甬道丹陛桥相连。祈谷坛西有斋宫。外坛西南分布有神乐署、牺牲所。它的面积比故宫(紫禁城)大得多,比颐和园小。天坛始建于明代公元 1420 年,为祭天之用。由于中国皇帝自称"天子",所以他们不敢使自己的住宅"紫禁城"比天的大。

天坛是现存世界上最大的祭天建筑群。1998 年 12 月被联合国教科文组织列入《世界遗产名录》。

祭天在中国有非常悠久的历史,自西周时期统治者提出"君权天授"的理论后,祭天作为统治者维护自己政权的一种活动,得到历代皇帝的重视。朱元璋在南京建立明朝后,立即在钟山之阳(南)建圜丘坛以祭天,在钟山之阴(北)建方泽坛以祭地。不久改天地分祭为合祭制度,建天地坛合祭皇天后土。

明成祖朱棣迁都北京后,仿照南京的规制建立了天地坛。北京天地坛建成于明朝永乐十八年十二月(公元 1421 年 2 月)主要建筑为大祀殿,基础在现在祈谷坛位置上,规模比南京天地坛大。

天地合祭 110 年后,嘉靖皇帝又将天地人祭改为分级。在北京城的东、南、西、北四郊建立了日坛、天坛、月坛和地坛,天坛又称为圜丘坛，专门祭祀皇天上帝。

天地分祀后大祀殿即废止。嘉靖十七年（公元1538年）拆大祀殿，在其旧址上仿古明堂建大享殿，以举行秋享上帝典礼。大享殿建在三层圆形坛基上，殿为圆形，三层重檐攒尖金顶。上檐蓝色，重檐黄色，下檐绿色，象征天地人三位一体。

清朝建立后，保留了明代的祭坛和祭祀制度，只是大享殿不再举行大享礼，改为举行祈谷典礼。乾隆十六年（公元1751年）将大享殿改名祈年殿，门改为祈年门，以符祈谷之意。次年三层重檐及皇乾殿、祈年门、皇穹宇等建筑的屋顶均改为蓝色琉璃瓦，以象天色。乾隆十四年（公元1749年），乾隆皇帝将圜丘扩建，坛制不变，但栏板、望柱改为汉白玉，坛面则改铺艾叶青石，上层中心为一圆形石块，外铺九圈扇形石块，内圈九块，其余以九的倍数递增，中、下层亦皆如此。各层栏板、望柱数目和台阶数都为九或九的倍数，以象征天数。北京天坛经过乾隆时期的调整完善之后，整个坛议制度最终形成。同时，祭天礼仪制度也在乾隆时期得以最后确定。

明清两朝，北京天坛共有22个皇帝举行过654次祭天大典。1911年清朝皇帝退位，祭天制度废除。天坛也从此失去了皇家祭坛的地位。但1914年袁世凯为复辟帝制，重新制定了一套祭天礼仪及祭祀服饰，于当年冬至举行祭天大典。不久袁世凯身亡，洪宪帝制被废。这次祭祀也成为天坛历史上最后一次祭天大典。

天坛被长长的围墙围着。墙里的北面部分是象征着上天的半圆环，南部是象征着大地的方形广场。北部高于南部。这种设计显示天在上，地在下，反映了中国古代人民"天圆地方"的思想。

两层围墙把天坛分为内坛和外坛。天坛的主体建筑位于内坛中轴线的南北两端。最宏伟的建筑物从南到北依次是圜丘坛、皇穹宇和祈年殿。还有其他的建筑物，如三音石和回音壁。

圜丘坛是三层汉白玉构成的平台。在明朝时期（公元1368年~公元1911年），皇帝将在每年的冬至祭天。此仪式是为了感谢上天并祈愿将来一切都好。祈年殿是一座大殿，圆顶，有三层屋檐。天坛另一个重要的建筑是皇穹宇。如果从远处看，你会发现皇穹宇像一把蓝伞。它的结构和祈年殿类似，但规模较小，由砖和木材构成，并被白色大理石栏杆围绕着。

皇穹宇殿前甬路从北面数，前三块石板即为"三音石"。当站在第一块石板上击一下掌，只能听见一声回音；当站在第二块石板上击一下掌就可以听见两声回音；当站在第三块石板上击一下掌便听到连续不断的三声回音。这就是为什么把这三块石板称为三音石的原因。另一个有趣的著名的地方是别具一格的回音壁。它围绕着皇穹宇，周长为193米。

作为中国祭天文化的物质载体，天坛积淀了深厚的民族文化内涵。天坛文化涉及中华民族文化的诸多领域，包括历史、政治、哲学、天文、建筑、历法、音乐、绘画、园林、伦理等等，是中国传统文化的集大成。

1998年12月2日，在日本京都召开的第22届世界遗产委员会会议上，天坛被列入世界遗产名录。同时，世界遗产委员会高度概括了天坛作为文化遗产的标准："一、天坛是建筑和景观设计之杰作，朴素而鲜明地体现出对世界伟大文明之一的发展产生过深刻影响的一种极其重要的宇宙观。二、许多世纪以来，天坛所独具的象征性布局和设计，对远东地区的建筑和规制产生了深刻的影响。三、两千多年来，中国一直处于封建王朝统治之下，而天坛的设计和布局正是这些封建王朝合法性的象征。"

目前，天坛公园正以其深厚的民族文化内涵、宏伟的古典建筑景观、古朴的祭坛环境氛围吸引着成千上万的中外游客前来参观游览。

（2）参考译文。

The Temple of Heaven

The Temple of Heaven is a worthwhile visiting place in Beijing. Located to the southeast of the Forbidden City, the Temple of Heaven was where emperors of the Ming (1368 – 1644) and Qing (1644 – 1911) Dynasties worshipped heaven. It covers 273 hectares and the layout is circular in the north and square in the south to symbolize the circular heaven and the square earth. The compound has two surrounding walls. The main buildings for worship are located within the inner wall. The main buildings – the Circular Mound Altar in the south and the Altar of Prayer for Good Harvestsin the north – are linked with a 360meter – long passage, called Danbiqiao or Red Stairway Bridge. To the west of the Altar of Prayer for Good Harvests lies the Hall of Abstinence; to the southwest are the Divine Music Administration and the Department for Sacrifices. It is much bigger than the Imperial Palace (the Forbidden City) and smaller than the Summer Palace. The temple was built in 1420 A. D. in the Ming Dynasty to offer sacrifice to Heaven. As Chinese emperors called themselves "the Son of Heaven", they dared not to build their own dwelling "the Forbidden City" bigger than the dwelling for Heaven.

The Temple of Heaven is the largest existing architectural complex in the world for the purpose of praying to heaven for good harvests. It was included in the World Heritage List by UNESCO in December 1998.

The Chinese have a long history of worshipping heaven. Because the rulers of the Western Zhou Dynasty (11th Century – 771 BC) claimed that they were empowered by heaven, the worship of heaven was continued by almost all state rulers as a way to uphold their right of administration. After Zhu Yuanzhang, the first emperor of the Ming Dynasty, established his rule in Nanjing, he ordered a circular mound altar built on the southern slope of Zhongshan Mountain to worship heaven and a square altar built on the northern slope to worship earth. Later the ceremonies to worship heaven and earth were merged and held in one temple.

After Zhu Di, the second emperor of the Ming Dynasty, moved the capital from Nanjing to Beijing, he had the Altar of Heaven and Earth built in the new capital based on the standards of the one in Nanjing, but larger in size. Construction was completed in February of 1421, at the site of today's Altar of Prayer for Good Harvests. The Temple of Worship was the main place for the ceremony.

About 110 years later, Emperor Jiajing issued an order to separate the worship ceremony of heaven from that of earth. Four altars were then built in the four directions (east, south, west and north) of the city of Beijing to worship the sun, heaven, the moon and the earth. The Temple of Heaven, also known as the Circular Mound Altar, was built to worship heaven.

The Hall of Worship was abandoned after the separation of the ceremonies and demolished in 1538. On that site was erected the Hall of Fruition, where ceremonies were held in autumn to present the harvest to heaven. The temple, sitting on a three-layer round base, was a round building with triple-eaved roof—the upper eave was blue, the middle eave yellow and the lower green. This symbolized the combination of the heaven, the earth and the people.

The rulers of the Qing Dynasty, which replaced the Ming Dynasty in 1644, kept the worship system, but changed the function of the Temple of Worship from presenting fruits to heaven to prayer for good harvests. The temple was renamed the temple of Prayer for Good Harvests in 1751, and the entrance was renamed the Gate of Prayer for Good Harvests. The following year the roofs of major buildings were replaced with blue glazed tiles to match the color of the sky. In 1749, Emperor Qianlong had the Circular Mound Altar expanded. The wooden railing and balustrades were replaced with white marble and the ground was paved with greenish stones. In the middle of the upper terrace was placed a round stone surrounded by nine concentric rings of paving stones. The number of stones in the first ring is nine, in the second 18, and so on, up to 81 in the ninth ring. The middle and bottom terraces are also nine rings each. Even the numbers of the carved railing, the balustrades and the steps are also in multiples of nine, to coincide with the astronomical phenomena. With all these readjustments, the altar system was finally completed and the worship of the heaven was reformed during the reign of Emperor Qianlong (1736 – 1795).

A total of 22 Ming and Qing emperors held 654 ceremonies to worship the heaven in Beijing. The worship was abandoned in 1911 when the last emperor of the Qing court abdicated and the Temple of Heaven ended its role as an imperial altar. However, the last ceremony held at the Temple was on the Winter Solstice in 1914 by Yuan Shikai who attempted to re-establish a monarchy. Yuan developed a new ceremony and tailored special attire therefore. Yuan died shortly thereafter and his regime was overthrown.

The Temple of Heaven is enclosed with a long wall. The northern part within the wall is semicircular symbolizing the heavens and the southern part is square symbolizing the earth. The northern part is higher than the southern part. This design shows that the heaven is high and the earth is low and reflects an ancient Chinese thought of "The heaven is round and the earth is square".

The Temple of Heaven is divided by two enclosed walls into inner part and outer part. The main buildings of the temple lie at the south and north ends of the middle axis line of the inner part. The most magnificent buildings from south to north are the Circular Terrace, the Imperial Heavenly Vault and the Hall of Prayer for Good Harvests. Also, there are some additional buildings like Three Echo Stones and Echo Wall.

Circular Mound Altar has three layered terraces with white marble. During the Ming and Qing Dynasties (1368A.D. – 1911A.D.), the emperors would offer sacrifices to Heaven on the day of the Winter Solstice every year. This ceremony was to thank Heaven and hope everything would be good in the future. Temple of Prayer for Good Harvests is a big palace with round roof and three layers of eaves. Another important building in the Temple of Heaven is Imperial Heavenly Vault. If you look at it from far away, you will find that the vault is like a blue umbrella. The structure of it is like that of the Temple of Prayer for Good Harvests, but smaller in size. The vault was made of bricks and timber, with white marble railings surrounded.

Three Echo Stones is outside of the gate of Imperial Heavenly Vault. If you speak facing the vault while standing on the first stone, you will hear one echo; standing on the second and the third stone, you will hear two and three echoes respectively. That's why these three stones are named the Three Echo Stones. Another interesting and famous place for you to visit is called Echo Wall owning special feature. The wall encloses Imperial Heavenly Vault. Its perimeter is 193 meters.

As the place for the worship ceremony, the Temple of Heaven has deep cultural connotations for the Chinese nation. It is a comprehensive expression of Chinese history and culture, politics, philosophy, astronomy, architecture, calendar system, music, painting, gardening and ethics. It synthesizes the traditional culture of the Chinese nation.

On December 2, 1998, the Temple of Heaven was placed onto the World Heritage List at the 22nd conference of the World Heritage Committee. The committee came to the conclusion that firstly, the Temple of Heaven is a masterpiece of architecture and landscape design which simply and graphically illustrates a cosmogony of great importance for the evolution of one of the world's great civilizations. Moreover, the symbolic layout and design of the Temple of Heaven had a profound influence on architecture and planning in the Far East over many centuries. Besides, for more than two thousand years China was ruled by a series of feudal dynasties, the legitimacy of which was symbolized by the design and layout of the Temple of Heaven.

Presently, the Temple of Heaven receives millions of visitors from near and afar who are impressed by the complicated cultural presentation, the grand, ancient architectural complex and the secluded surroundings.

（3）建筑专业术语及词汇。

① 天坛 the Temple of Heaven

② 祭天大典 worshipping heaven ceremony

③ 平面布局 layout

④ 天圆地方 the round heaven and the square earth

⑤ 甬道 passage

⑥ 祭天建筑群 architectural complex for the purpose of praying to heaven for good harvests

⑦ 三层重檐顶 triple-eaved roof

⑧ 天地人三位一体 a combination of the heaven, earth and the people

⑨ 皇家祭坛 imperial altar

⑩ 祭祀制度 worship system

⑪ 蓝色琉璃瓦 blue glazed tiles

⑫ 汉白玉 white marble

⑬ 艾叶青石 greenish blue stone

⑭ 景观设计 landscape design

⑮ 封建王朝 feudal dynasties

⑯ 栏板 carved railing

⑰ 望柱 balustrades

⑱ 圆丘坛 the Circular Terrace

⑲ 皇穹宇 the Imperial Heavenly Vault

⑳ 祈年殿 the Hall of Prayer for Good Harvests.

㉑ 公顷 hectare

㉒ 场域 compound

㉓ 住处，居所 dwelling

㉔ 授权，赋予权力 empower

㉕ 维护政权 uphold the right of administration

㉖ 斜坡 slope

㉗ 合并，融合 merge

㉘ 废弃，放弃 abandon

㉙ 废除，推翻 demolish

㉚ 建造，竖立 erect

㉛ 同心圆 concentric rings

㉜ 天数 astronomical phenomena

㉝ 调整 readjustment

㉞ 退位 abdicate

㉟ 帝制 monarchy

㊱ 特别订制服饰 tailor special attire

㊲ 冬至 the Winter Solstice

㊳ 砖和木 brick and timber

㊴ 政权，政体 regime

㊵ 推翻 overthrow

㊶ 附上，围着 be enclosed with

㊷ 半圆环的 semicircular

㊸ 回音 echo

㊹ 周长 perimeter

㊺ 杰作 masterpiece

㊻ 生动地 graphically

㊼ 世界观 cosmogony

㊽ 合法性 legitimacy

㊾ 隐蔽的 secluded

㊿ 文化内涵 cultural connotation

4.4.2.2　Text B 外国建筑——The Empire State Building

（1）文本原文。

帝国大厦

帝国大厦，是位于美国纽约州纽约市曼哈顿第五大道350号、西33街与西34街之间的一栋著名摩天大楼，名称源于纽约州的昵称——帝国州，故其英文名称原意为纽约州大厦或者帝国州大厦。惟帝国大厦的翻译已经约定俗世，及沿用至今。帝国大厦为纽约市以至美国最著名的地标和旅游景点之一，为美国及美洲第4高，世界上第25高的摩天大楼，也是保持世界最高建筑地位最久的摩天大楼（1931~1972年）。

18世纪后期，此地曾经是一个农场。19世纪后期，此地变为社会名流经常光顾的华道夫－阿斯多里亚酒店。1930年1月22日，项目开始动工，于同年3月17日开始建筑。项目涉及了3400名工人，主要是欧洲移民，也包括数百名蒙特利尔附近的北美原住民。据官方统计，施工过程中共有5名工人身亡。帝国大厦的建设速度是每星期建4层半。在当时的技术水平下，是惊人的。整座大厦最后提前了5个月落成启用，比预计的五千万美元减少了10%，所用材料包括5660立方米的印第安那州石灰岩和花岗岩，1000万块砖和730吨铝和不锈钢。

1931年5月1日，美国总统赫伯特·胡佛在首都华盛顿特区亲自按下电钮，点亮大厦灯光，帝国大厦正式落成，成为当时世界最高的建筑。但许多办公室在40年代之前一直空置，使它在早期被戏称为"空国大厦"。

2001年，"9·11"事件发生后，人们曾一度担心帝国大厦是否会成为恐怖袭击的下一个目标。在经历了短暂的关闭之后，帝国大厦第86层的观景平台重新对公众开放，只不过为了防止有人从该处跳楼，观景台周围的防护铁栏又加固了而已。观景平台位于大楼1050英尺处（约320米），从该处可看到纽约市的全貌。

2006年5月1日，美国纽约的帝国大厦度过了它的75岁"生日"，美国的报纸和电视为此做了许多报道，同时也追忆大厦经历过的沧桑。

2008年2月，由克林顿气候行动计划（The Clinton Climate Initiative，简称CCI）发起的帝国大厦减排项目悄然启动。一年半之后，帝国大厦改造完成，比改造前减少38%的二氧化碳排放量。

20世纪30年代，美国经济处于大萧条时期，人民生活更加困苦，而华尔街的老板们却热衷于修建摩天大楼的竞赛。百万富翁拉斯科布为了显示自己的富有，决意修建一座世界最高的大厦。他找来著名的建筑师威廉·拉姆，问大厦能盖多高，拉姆沉思片刻后回答："1050英尺"。拉斯科布对这个高度很不满意，因为这仅仅比当时纽约新建成的克莱斯勒大厦高4英尺。于是，建筑师设法增加了一节200英尺高的圆塔，使帝国大厦的高度为1250英尺。这座摩天大楼只用了410天就建成，算是建筑史上的奇迹。最初的计划是建一幢看上去低矮结实的34层大厦，后来又作过16次修改，最后才采纳了拉斯科布的"铅笔型"方案。最终采取的方案使得建筑十分牢固。

帝国大厦原本共381米，20世纪50年代安装的天线使它的高度上升至443.7米。根据估算，建造帝国大厦的材料约有330000吨。大厦总共拥有6500个窗户、73部电梯，从底层步行至顶层须经过1860级台阶，总建筑面积为204385平方米。

帝国大厦里面的墙壁装饰多为来自意大利、法国、比利时、德国的不同颜色的大理石，一楼大厅是各种艺术品的殿堂。自1964年起，大厦上部30层的外表全部用彩灯装饰，通宵闪亮。在大厦上最早安装的灯是1932年美国总统大选时的一架探照灯，意在向方圆80公里的民众宣告小罗斯福成功当选新一届美国总统。1956年，被称为自由之光的旋转灯安装到大厦顶部。1984年，自动变色灯装上了大厦顶端，灯光的表现力变得更为丰富多彩。由于纽约市华人、华侨众多，从2001年开始，帝国大厦会在每年春节、每年中华人民共和国国庆节期间晚上点亮象征吉祥的红、黄两色彩灯。在"9·11事件"后大厦也连续3个月点亮蓝色灯光以示哀悼。

（2）参考译文。

The Empire State Building（Empire State Building）is a famous skyscraper between Fifth Avenue 350, West 33 street and West 34 street in Manhattan, New York, United States. The name comes from the nickname of New York, the Empire State, so its English name is originally intended for the New York state building or the Empire state building. However, the translation of The Empire State Building has been stipulated in the world and has been used till now. The Empire State Building is one of the most famous landmarks and tourist attractions in New York and the United States. It is fourth high in the United States and America, the twenty-fifth tallest skyscraper in the world, and is also the oldest skyscraper in the world（1931-1972 years）.

In the late eighteenth Century, this land was once a farm. In the late nineteenth Century, this place was once a hotel frequented by the celebrities of Waldorf Astoria. In January 22, 1930, the project started construction and began construction in March 17th of the same year. The project involved 3400 workers, mainly European immigrants, including hundreds of North American Aborigines near Montreal. According to official statistics, 5 workers died during the construction. The Empire State Building is built at 4 and a half times a week. At the level of technology at that time, it was amazing. The building ended 5 months in advance, reduced by 10% than the estimated 50 million dollars, including 5660 cubic meters of Indiana limestone and granite, 10 million bricks and 730 tons of aluminum and stainless steel.

On May 1st, 1931, the president of the United States, Herbert Huff, pressed the button in the capital, Washington, D. C. , lit the lighting of the building, and the Empire State Building was formally established as the highest building in the world at that time. But many offices were vacant until 40s, making it known as Empty State Building in the early days.

In 2001, after the "9 · 11" incident, people once worried that the Empire State Building would become the next target of terrorist attacks. After a brief closure, the eighty – sixth floor view platform of the Empire State Building was reopened to the public, only to prevent someone from jumping from the floor, and the protective iron fence around the observatory was reinforced. The viewing platform is located at 1050 feet (about 320 meters) of the building, and the whole picture of New York can be seen from it.

On May 1st, 2006, The Empire State Building in New York, USA, spent its 75 – year – old "birthday", and American newspapers and television had done a lot of reports and recalled the vicissitudes that the building had experienced.

In February 2008, the Empire State Building emission reduction project launched by the Clinton Climate Initiative (CCI) quietly started. A year and a half later, The Empire State Building was completed, reducing carbon dioxide emissions by 38%.

In 1930s, when the American economy was in the great depression, people's lives became more and more difficult, while Wall Street bosses were keen on building skyscrapers. Millionaire Raskob, in order to show his wealth, is determined to build the tallest building in the world. He found the famous architect William Ram and asked how tall the building could be. After a moment of contemplation, Rahm replied, "1050 feet". Raskob was not satisfied with the height, because it was only 4 feet higher than the new Chrysler mansion in New York. So the architect managed to add a 200 foot high Round Tower to the height of the Empire State Building 1250 feet. The skyscraper was built in just 410 days, a miracle in the history of architecture. The initial plan was to build a low and sturdy 34 story building, and later made 16 revisions, and finally adopted the "pencil type" scheme of the rich Raskob. The final plan made the building very strong.

The Empire State Building originally had a total of 381 meters, and the antenna installed in 1950s raised its height to 443.7 meters. According to the estimate, there are about 330000 tons of material for The Empire State Building. The building has 6500 windows and 73 elevators. The 1860 steps are required to walk from the bottom to the top, with a total construction area of 204385 square meters.

The walls in The Empire State Building are decorated with different colors of marble from Italy, France, Belgium and Germany, and the hall on the first floor is the palace of various works of art. Since 1964, the 30 upper floors of the building have been decorated with colored lanterns, shining all night. The first lamp installed on the building was a searchlight in the 1932 presidential election in the United States, intended to announce that little Roosevelt was elected to the new president of the United States within 80 kilometers to the square. In 1956, the revolving light, called freedom light, was installed at the top of the building. In 1984, the automatic colorful lamp was installed on the top of the building, and the expressive power of the lights became more colorful. Since the number of Chinese and overseas Chinese in New York, from 2001, the Empire State Building will light up the red and yellow two color lights, which symbolize the happy and auspicious, during the annual Spring Festival and the national day of People's Republic of China. After the 911 incident, the building also lit up the blue lights for 3 consecutive months in order to express mourning.

(3) Terms and Vocabularies。

① the Fifth Avenue 第五大道

② tourist attractions 旅游景点

③ limestone 石灰岩

④ granite 花岗岩

⑤ bricks 砖

⑥ aluminum 铝

⑦ stainless steel 不锈钢

⑧ marble 大理石

⑨ view platform 观景平台

⑩ protective iron fence 防护铁栏

⑪ reinforce 加固

⑫ press the button 按下电钮

⑬ carbon dioxide emissions 二氧化碳排放量

⑭ emission reduction project 减排项目

⑮ construction area 建筑面积

⑯ decorate 装饰

⑰ colored lanterns 彩灯

⑱ the palace of various works of art 艺术品的殿堂

⑲ searchlight 探照灯

⑳ skyscraper 摩天大楼

㉑ landmark 地标

㉒ observatory 天文台，瞭望台

㉓ expressive power 表现力

㉔ marble 大理石

㉕ stipulate 规定，保证

㉖ celebrity 名人

㉗ immigrant 移民

㉘ vacant 空闲的，空缺的

㉙ terrorist attack 恐怖袭击

㉚ vicissitudes 变迁，兴衰

㉛ closure 终止，结束

㉜ emission reduction project 减排项目

㉝ carbon dioxide emissions 二氧化碳排放

㉞ contemplation 沉思，冥想

㉟ miracle 奇迹

㊱ initial 最初的，初始的

㊲ sturdy 坚定的

㊳ scheme 计划，方案

㊴ presidential election 总统选举

㊵ antenna 天线

㊶ auspicious 吉祥的

㊷ consecutive 连续的

㊸ mourning 哀痛

㊹ the revolving light 旋转灯

㊺ automatic colorful lamp 自动变色

㊻ revision 修改，修订

㊼ the great depression 大萧条

㊽ architect 建筑师

㊾ adopt "pencil type" 采用铅笔型

㊿ freedom light 自由之光

4.4.3 翻译练习

4.4.3.1 句子翻译

1. 汉译英

① 天坛位于北京紫禁城的东南方，是明、清两朝皇帝举行祭天大典的场所。

② 天坛始建于明代公元 1420 年，为祭天之用。

③ 天坛是现存世界上最大的祭天建筑群。

④ 祭天在中国有非常悠久的历史。

⑤ 不久改天地分祭为合祭制度，建天地坛合祭皇天后土。

⑥ 嘉靖十七年（公元 1538 年）拆大祀殿，在其旧址上仿古明堂建大享殿，以举行秋享上帝典礼。

⑦ 清朝建立后，保留了明代的祭坛和祭祀制度，只是大享殿不再举行大享礼，改为举行祈谷典礼。

⑧ 北京天坛经过乾隆时期的调整完善之后，整个坛议制度最终形成。同时，祭天礼仪制度也在乾隆时期得以最后确定。

⑨ 1911 年清朝皇帝退位，祭天制度废除。天坛也从此失去了皇家祭坛的地位。

⑩ 天坛被长长的围墙围着。墙里的北面部分是象征着上天的半圆环，南部是象征着大地的方形广场。

⑪ 两层围墙把天坛分为内坛和外坛。

⑫ 在明朝时期（公元 1368 年～公元 1911 年），皇帝将在每年的冬至祭天。

⑬ 此仪式是为了感谢上天并祈愿将来一切都好。

⑭ 作为中国祭天文化的物质载体，天坛积淀了深厚的民族文化

内涵。

⑮ 同时，世界遗产委员会高度概括了天坛作为文化遗产的标准："一、天坛是建筑和景观设计之杰作，朴素而鲜明地体现出对世界伟大文明之一的发展产生过深刻影响的一种极其重要的宇宙观。"

⑯ 许多世纪以来，天坛所独具的象征性布局和设计，对远东地区的建筑和规制产生了深刻的影响。

⑰ 两千多年来，中国一直处于封建王朝统治之下，而天坛的设计和布局正是这些封建王朝合法性的象征。"

⑱ 目前，天坛公园正以其深厚的民族文化内涵、宏伟的古典建筑景观、古朴的祭坛环境氛围吸引着成千上万的中外游客前来参观游览。

⑲ 天坛文化涉及中华民族文化的诸多领域，包括历史、政治、哲学、天文、建筑、历法、音乐、绘画、园林、伦理等等，是中国传统文化的集大成。

⑳ 它的结构和祈年殿类似，但规模较小，由砖和木材构成，并被白色大理石栏杆围绕着。

2. 英译汉

① The Empire State Building（Empire State Building）is a famous skyscraper between Fifth Avenue 350, West 33 street and West 34 street in Manhattan, New York, United States.

② However, the translation of The Empire State Building has been stipulated in the world and has been used till now.

③ The Empire State Building is one of the most famous landmarks and tourist attractions inNew York and the United States.

④ The Empire State Building is built at 4 and a half times a week.

⑤ The building ended 5 months in advance, reduced by 10% than the estimated 50 million dollars, including 5660 cubic meters of Indiana limestone and granite, 10 million bricks and 730 tons of aluminum and stainless steel.

⑥ On May 1st, 1931, the president of the United States, Herbert Huff, pressed the button in the capital, Washington, D. C. , lit the lighting of the building, and the Empire State Building was formally established as the highest building in the world at that time.

⑦ But many offices were vacant until 40s, making it known as Empty State Building in the early days.

⑧ After a brief closure, the eighty – sixth floor view platform of the Empire State Building was reopened to the public, only to prevent someone from jumping from the floor, and the protective iron fence around the observatory was reinforced.

⑨ On May 1st, 2006, The Empire State Building in New York, USA, spent its 75 – year – old "birthday", and American newspapers and television had done a lot of reports and recalled the vicissitudes that the building had experienced.

⑩ In February 2008, the Empire State Building emission reduction project launched by the Clinton Climate Initiative (CCI) quietly started.

⑪ A year and a half later, The Empire State Building was completed, reducing carbon dioxide emissions by 38%.

⑫ In 1930s, when the American economy was in the great depression, people's lives became more and more difficult, while Wall Street bosses were keen on building skyscrapers.

⑬ The skyscraper was built in just 410 days, a miracle in the history of architecture. The initial plan was to build a low and sturdy 34 story building, and later made 16 revisions, and finally adopted the "pencil type" scheme of the rich Raskob.

⑭ The Empire State Building originally had a total of 381 meters, and the antenna installed in 1950s raised its height to 443. 7 meters.

⑮ The 1860 steps are required to walk from the bottom to the top, with a total construction area of 204385 square meters.

⑯ The walls in the Empire State Building are decorated with different colors of marble from Italy, France, Belgium and Germany, and the hall on the first floor is the palace of various works of art.

⑰ The first lamp installed on the building was a searchlight in the 1932 presidential election in the United States, intended to announce that little Roosevelt was elected to the new president of the United States within 80 kilometers to the square.

⑱ In 1984, the automatic colorful lamp was installed on the top of the building, and the expressive power of the lights became more colorful.

⑲ Since the number of Chinese and overseas Chinese in New York, from 2001, the Empire State Building will light up the red and yellow two color lights, which symbolize the happy and auspicious, during the annual Spring Festival and the national day of People's Republic of China.

⑳ After the 911 incident, the building also lit up the blue lights for 3 consecutive months in order to express mourning.

4.4.3.2 段落翻译

1. 汉译英

① 中国园林是经过三千多年演化而形成的独具一格的园林景观，是中国建筑不可分割的一部分。从建园者身份来看，中国园林主要分为两类：皇家园林和私家园林。皇家园林式为皇室成员享乐而建造，大多数皇家园林在北方，最著名的有北京颐和园和河北承德的避暑山庄。私家园林主要由学者、商人和官员为摆脱嘈杂的外部世界儿建造，南方比较多见，如苏州的拙政园和留园。典型的中国园林周围有围墙，园内有池塘、假山、树木、花草以及各种各样由弯曲的小路和走廊衔接的建筑，构成了一种人与自然和谐相处的微缩景观。传统建筑注重对称美，而园林往往讲究尽量不对称，以忠实再现自然美。

② 岳阳楼位于湖南省岳阳市洞庭湖畔，是古代建筑的瑰宝。岳阳楼与黄鹤楼、滕王阁并称为"江南三大名楼"，有"洞庭天下水，岳阳天下楼"的美誉。岳阳楼是一座三层长方形建筑，完全由木料建成。一般认为岳阳楼建于1,700多年前的三国时期。后毁于战乱，宋朝时重建，从此成为旅游景点。但岳阳楼真正名闻天下，是在北宋藤子京重建、范仲淹作记之后。岳阳楼是"江南三大名楼"中唯一保持原貌的古建筑，具有极高的建筑艺术价值。1988年岳阳楼被国务院列为国家重点文物保护单位。

③ 福建土楼是中国福建西南山区的一种独特的居住性建筑。由于大多数土楼为客家人所建，福建土楼是客家文化的一种象征，故又称"客家土楼"。土楼产生于宋元，成熟于明末、清代和民国时期。土楼是一种坚固的土建筑，通常呈圆形或长方形，其建筑材料主要是土、木、石、竹。在布局上，土楼与周围的山水融合在一起，反映了中国传统建筑的设计理念。同时，土楼采光好，通风好，防风抗震，冬暖夏凉，是世界上独一无二的大型夯土居民建筑，是建筑艺术的杰作。2008年，有46座福建土楼被联合国教科文组织列入世界遗产名录。

2. 英译汉

① The Louvre located on the right bank of the Seine and in the center of Paris, is one of the world's biggest museums, famous for its collection of classical paintings and sculptures. Construction of the Louvre began in late 12th century, originally built as a defensive fortress, and later as a royal palace until Louis XIV chose Versailles as his residence in 1682. The building was extended many times to form the present Louvre Palace. In 1793, the Louvre was converted into a national art museum and opened to the public the same year. In 1981, French President Francois Mitterrand announced a plan to restore the Louvre Palace. The pyramid entrance to the museum by the Chinese – American architect Leoh Ming Pei adds to the glamor of the Louvre and becomes a new shining point of the museum.

② Angkor Wat is a Buddhist temple complex in Cambodia and the largest religious monument in the world. As the best-preserved temple, it is the only one to have remained a significant religious center since its foundation. Angkor Wat is one of the most beautiful and most mysterious historic sites in the world. It has become a symbol and a source of great national pride of Cambodia. Since the 1990s, Angkor Wat has seen continued conservation efforts and a massive increase in tourism. As the country's prime attraction for visitors, it is drawing more and more tourists from around the world. In 1992, the UNESCO World Heritage Committee declared the monument a World Heritage Site.

③ The Colosseum is a vast elliptical amphitheater of ancient Rome. Made of concrete and stone, it is located east of the Roman Forum, in the center of Italy. As the largest amphitheater in the world, it is a vast ellipse of over 500 meters in perimeter. The Colosseum has four stories and each of the first three stories has 80 arched entrance, which enabled spectators to enter and exit in good order. The central oval area is able to accommodate some 50000 spectators who may have been arranged according to social ranking. The arena was designed mainly for gladiatorial contests and other public activities. The Colosseum is deemed to be one of the greatest works of Roman architecture and engineering and an iconic symbol of Imperial Roman. Modem sports stadiums are influenced by the design of the Colosseum in one way or another.

4.4.4 练习答案

4.4.4.1 句子翻译

1. 汉译英答案

① Located to the southeast of the Forbidden City, the Temple of Heaven was where emperors of the Ming (1368A. D. – 1644A. D.) and Qing (1644 – 1911) Dynasties worshipped heaven.

② The temple was built in 1420 A. D. in the Ming Dynasty to offer sacrifice to Heaven.

③ The Temple of Heaven is the largest existing architectural complex in the world for the purpose of praying to heaven for good harvests.

④ The Chinese have a long history of worshipping heaven.

⑤ Later the ceremonies to worship heaven and earth were merged and held in one temple.

⑥ On that site was erected the Hall of Fruition, where ceremonies were held in autumn to present the harvest to heaven.

⑦ The rulers of the Qing Dynasty, which replaced the Ming Dynasty in 1644, kept the worship system, but changed the function of the Temple of Worship from presenting fruits to heaven to prayer for good harvests.

⑧ With all these readjustments, the altar system was finally completed and the worship of the heaven was reformed during the reign of Emperor Qianlong (1736 – 1795).

⑨ The whole building space of theImperial Palace is rich in space, magnificent in size and magnificent in appearance. It shows the grand momentum of solemn monarch.

⑩ The Temple of Heaven is enclosed with a long wall. The northern part within the wall is semicircular symbolizing the heavens and the southern part is square symbolizing the earth.

⑪ The Temple of Heaven is divided by two enclosed walls into inner part and outer part.

⑫ During the Ming and Qing Dynasties (1368A. D. – 1911A. D.), the emperors would offer sacrifices to Heaven on the day of the Winter Solstice every year.

⑬ This ceremony was to thank Heaven and hope everything would be good in the future.

⑭ As the place for the worship ceremony, the Temple of Heaven has deep cultural connotations for the Chinese nation.

⑮ The committee came to the conclusion that firstly, the Temple of Heaven is a masterpiece of architecture and landscape design which simply and graphically illustrates a cosmogony of great importance for the evolution of one of the world's great civilizations.

⑯ Moreover, the symbolic layout and design of the Temple of Heaven had a profound influence on architecture and planning in the Far East over many centuries.

⑰ Besides, for more than two thousand years China was ruled by a series of feudal dynasties, the legitimacy of which was symbolized by the design and layout of the Temple of Heaven.

⑱ Presently, the Temple of Heaven receives millions of visitors from near and far who are impressed by the complicated cultural presentation, the grand, ancient architectural complex and the secluded surroundings.

⑲ It is a comprehensive expression of Chinese history and culture, politics, philosophy, astronomy, architecture, calendar system, music, painting, gardening and ethics.

⑳ The structure of it is like that of the Temple of Prayer for Good Harvests, but smaller in size. The vault was made of bricks and timber, with white marble railings surrounded.

2. 英译汉答案

① 帝国大厦，是位于美国纽约州纽约市曼哈顿第五大道350号、西33街与西34街之间的一栋著名摩天大楼。

② 惟帝国大厦的翻译已经约定俗世，及沿用至今。

③ 帝国大厦为纽约市以至美国最著名的地标和旅游景点之一。

④ 帝国大厦的建设速度是每星期建 4 层半。

⑤ 整座大厦最后提前了 5 个月落成启用，比预计的 5 千万美元减少了 10%，所用材料包括 5660 立方米的印第安那州石灰岩和花岗岩，1000 万块砖和 730 吨铝和不锈钢。

⑥ 1931 年 5 月 1 日，美国总统赫伯特·胡佛在首都华盛顿特区亲自按下电钮，点亮大厦灯光，帝国大厦正式落成，成为当时世界最高的建筑。

⑦ 但许多办公室在 40 年代之前一直空置，使它在早期被戏称为"空国大厦"。

⑧ 在经历了短暂的关闭之后，帝国大厦第 86 层的观景平台重新对公众开放，只不过为了防止有人从该处跳楼，观景台周围的防护铁栏又加固了而已。

⑨ 2006 年 5 月 1 日，美国纽约的帝国大厦度过了它的 75 岁"生日"，美国的报纸和电视为此做了许多报道，同时也追忆大厦经历过的沧桑。

⑩ 2008 年 2 月，由克林顿气候行动计划（The Clinton Climate Initiative，简称 CCI）发起的帝国大厦减排项目悄然启动。

⑪ 一年半之后，帝国大厦改造完成，比改造前减少 38% 的二氧化碳排放量。

⑫ 20 世纪 30 年代，美国经济处于大萧条时期，人民生活更加困苦，而华尔街的老板们却热衷于修建摩天大楼的竞赛。

⑬ 这座摩天大楼只用了 410 天就建成，算是建筑史上的奇迹。最初的计划是建一幢看上去低矮结实的 34 层大厦，后来又作过 16 次修改，最后才采纳了拉斯科布的"铅笔型"方案。

⑭ 帝国大厦原本共 381 米，20 世纪 50 年代安装的天线使它的高度上升至 443.7 米。

⑮ 大厦总共拥有 6500 个窗户、73 部电梯，从底层步行至顶层须经过 1860 级台阶，总建筑面积为 204385 平方米。

⑯帝国大厦里面的墙壁装饰多为来自意大利、法国、比利时、德国的不同颜色的大理石,一楼大厅是各种艺术品的殿堂。

⑰在大厦上最早安装的灯是1932年美国总统大选时的一架探照灯,意在向方圆80公里的民众宣告小罗斯福成功当选新一届美国总统。

⑱1984年,自动变色灯装上了大厦顶端,灯光的表现力变得更为丰富多彩。

⑲由于纽约市华人、华侨众多,从2001年开始,帝国大厦会在每年春节、每年中华人民共和国国庆节期间晚上点亮象征吉祥的红、黄两色彩灯。

⑳在"9·11"事件后大厦也连续3个月点亮蓝色灯光以示哀悼。

4.4.4.2 段落翻译

1. 汉译英答案

① Classical Chinese gardens are a unique landscape garden style which has developed over 3000 years. It is an integral part of Chinese architecture. In terms of garden owners, classical Chinese gardens can be classified into two types: imperial gardens and private gardens. Imperial gardens are built for pleasure of the royal members. Most imperial gardens are located in north China, among which the Summer Palace in Beijing and the Chengde Mountain Resort in Chengde, Hebei province are most well-known. Private gardens are most built by scholars, merchants, or government officials attempting to seek a shelter from the bustling outside world. Private gardens are mainly found in south China, such as the Humble Administrator's Garden and the Lingering Garden in Suzhou. A typical Chinese garden is surrounded by walls and consists of ponds, rock works, plants and a variety of buildings connected by zig-zag paths and winding corridors, creating an idealized landscape in miniature, showing the harmony between man and nature. Unlike traditional architecture with emphasis on symmetry, classical Chinese gardens tend to be as asymmetrical as possible to reproduce faithfully the beauty of nature.

② The Yueyang Tower, an ancient architectural treasure, is located the shore of Dongting Lake in Yueyang City, Hunan Province. Alongside the Yellow Crane Tower and the Pavilion of Prince Teng, it is one of the "three great towers in the south of Yangtze River", enjoying the reputation of "Dongting Lake represents the most beautiful water while Yueyang Tower represents the most beautiful tower under heaven". The tower is a three-story rectangular building and is constructed entirely from wood. It is generally believed that it was built more than 1,700 years ago during the Three Kingdoms Period. The Tower was later damaged in the chaos of wars and was rebuilt in the Song Dynasty. Since then, it was become a tourist attraction. But the Tower became world-famous only when Teng Zijing rebuilt the Tower and Fan Zhongyan wrote the famous Yueyang Lou Ji (Memorial to Yueyang Tower) in the Northern Song Dynasty. The Yueyang Tower is the only one among the "three great towers" that keeps its original look, enjoying great architectural art value. In the 1988 the Tower was designated by the State Council as major historical and cultural site under state protection.

③ Fujian Tulou is a unique residential architecture in the mountainous areas in the southeastern Fujian of China. As most Tulou buildings were built by Hakka pople ofFujian Province, Fujian Tulou becomes a symbol of Hakka culture and is also called "Hakka Tulou". Tulou first appeared in the Song and Yuan dynasties and reached its maturity during the late Ming Dynasty, the Qing dynasty and the period of the Republic of China. Tulou is a fortified earth building, most commonly circular or rectangular in configuration, with earth, wood, stone and bamboo as the main building material. In layout, Tulou buildings are harmoniously integrated with the surrounding hills and waters, reflecting the concept of traditional Chinese architectural designs. Meanwhile, Tulou is a will‐lit, well‐ventilated, windproof and and earthquake‐proof building that is warm in winter and cool in summer. As a kind of large civilian residence of rammed earth, Tulou is found nowhere else in the world. It is a great work of architectural art. In 2008, a total of 46 Fujian Tulou sites wre inscribed by UNESCO o the World Heritage List.

2. 英译汉答案

① 卢浮宫位于巴黎市中心，赛纳河右岸，是世界上最大的博物馆之一，以收藏丰富的古典绘画和雕刻作品而闻名于世。卢浮宫始建于12世纪末，最初是用作防御性的城堡，后来成为法国的王宫，直到1682年路易十四将王宫搬往凡尔赛宫，卢浮宫经历多次扩建，才达到今天的规模。1793年，卢浮宫被辟为国立美术博物馆，同年向公众开放。1981年，法国总统弗朗索瓦·密特朗宣布了修整卢浮宫的计划。由华裔美籍设计师贝聿铭设计的金字塔型入口为卢浮宫锦上添花，成为一个新的亮点。

② 吴哥窟是柬埔寨的一处佛教庙宇群，也是世界上最大的宗教遗址。他是世界上保存最完整的庙宇，也是唯一一个自建寺以来始终具有重要地位的宗教中心。作为世界上最美丽神奇的古迹之一，吴哥窟已经成为柬埔寨的标志，让柬埔寨人民感到无比骄傲。20世纪90年代以来，针对吴哥窟的保护措施一直没有间断，游客人数迅猛增加。作为柬埔寨的主要旅游点，吴哥窟正吸引着越来越多的外国游人。1992年，联合国教科文组织世界遗产委员会将吴哥窟列入了世界遗产名录。

③ 罗马竞技场是古罗马时期的一所巨大的椭圆形露天竞技场，位于意大利中心古罗马广场东面，以混凝土和石头筑成，罗纳竞技场是世界上最大规模的露天竞技场，外观呈椭圆形，圆周长超过五百米，罗马竞技场有四层，下面三层每层都有80个拱形入口可供观众

井然有序地进出竞技场。中部的椭圆形竞技场能容纳约五万名观众，坐席根据不同社会阶层安排。竞技场主要是用来角斗和其他公共活动。罗马竞技场被认为是罗马建筑工程的伟大杰作，是罗马帝国的标志与象征。现代体育场都或多或少都受到罗马竞技场设计的影响。

4.5 翻译技巧：名词性从句的译法

建筑文本：中国建筑——苏州园林

外国建筑——英国议会大厦

4.5.1 翻译技巧——名词从句的译法

英语名词从句包括主语从句、宾语从句、表语从句和同位语从句等。翻译这类从句时，大多数可按原文的句序译成对应的汉语，但也还有一些其他处理方法。现分述如下。

1. 主语从句

（1）以 what, whatever, whoever 等代词引导的主语从句翻译时一般可按原文顺序翻译。

① What he told me was only half‐truth.

他告诉我的只是些半真半假的东西而已。

② Whatever he saw and heard on his trip gave him a very deep impression.

他此行所见所闻都给他留下了深刻的印象。

（2）以 it 作假主语所引出的真主语从句，翻译时视情况可以提前，也可以不提前。

其一，真主语从句提前译，为了强调起见，it 一般可以译出来，如果不需要强调，it 也可以不译出来。

① It doesn't make much difference whether he attends the meeting or not.

他参加不参加会议没有多大关系。

② It is a fact that the U.S.S.R. has sent its fleet to all parts of the world.

苏联已把它的舰队派往世界各地，这是事实。

其二，真主语从句不提前，it 一般不需要译出来。

① It is strange that she should have failed to see her own shortcomings.

真奇怪，她竟然没有看出自己的缺点。

② It was obvious that I had become the pawn in some sort of toplevel power play.

很清楚，某些高级人物在玩弄权术，而我却成了他们的工具。

2. 宾语从句

用 that，what，how 等引起的宾语从句汉译时一般不需要改变它在原句中的顺序。有时可加"说"字，再接下去译原文宾语从句的内容。

① I told him that because of the last condition, I'd have to turn it down.

我告诉他，由于那最后一个条件，我只得谢绝。

② He would remind people again that it was decided not only by himself but by lots of others.

他再三提醒大家说，决定这件事的不只是他一个人，还有其他许多人。

3. 表语从句

英语表语从句跟宾语从句一样，一般可按原文顺序翻译。

① His view of the press was that the reporters were either for him or against him.

他对新闻界的看法是，记者们不是支持他，就是反对他。

② This is what he is eager to do.

这就是他所渴望做的。

4. 同位语从句

同位语是用来对名词（或代词）作进一步的解释。同位语可以是单词，短语或从句，这里先介绍英语同位语从句的译法。

（1）同位语从句汉译时不提前。

① He expressed the hope that he would come over to visit China again.

他表示希望再到中国来访问。

② There were also indications that intelligence, not politics was Helms' primary concern.

而且也有种种迹象表时，赫尔姆斯主要关心的是情报，而不是政治。

（2）同位语从句提前。

① Yet, from the beginning, the fact that I was alive was ignored.

然而，从一开始，我仍然活着这个事实却偏偏被忽视了。

② But I knew I couldn't trust him. There was always the possibility that he was a political swindler.

但我知道不能轻信他。他是政治骗子这种可能性总是存在的。

（3）增加"即"（或"以为"）或用冒号，破折号分开。

① "Influenced by these ethics, Powers lived under the delusion that money does not stink…"

"受了这种道德观念的熏陶，鲍尔斯生活在一种错觉中，以为金钱总是香喷喷的……"

② But considered realistically, we had to face the fact that our prospects were less than good.

但是现实地考虑一下，我们不得不正视这样的事实：我们的前景并不妙。

例句——汉译英

① 一般认为岳阳楼建于1700多年前的三国时期。

It is generally believed that it was built more than 1700 years ago during the Three Kingdoms Period.

② 竞技场主要是用来进行角斗和举行其他公共活动。

The northern section, or theInner Court, was where the emperor lived with his royal family.

③ 原因是当时人们认为黑色代表水，可以灭火。

The reason is that it was believed black represented water and could extinguish fire.

④ 中国有句俗话说，"五岳归来不看山，黄山归来不看岳。"

A Chinese saying goes that "You won't want to visit any other mountains after seeing Wu Yue, but you won't wish to see even Wu Yue after returning from Mount Huangshan."

⑤ 可见黄山之美与五月相比由有过之而无不及。

It can be seen that the beauty of Mount Huangshan eclipses even Wu Yue.

例句——英译汉

① But some people have speculated that it was a temple built for the worship of ancient earth deities.

但是有些人推测，它是古时为供奉地球上的神灵而建造的庙宇。

② There is also archaeological evidence that was once used as a burial ground from its earliest beginning.

也有考古发现表明，巨石阵从建造出其就曾作为墓地使用。

③ Whatever its purpose, the astonishing scale, the beauty of the stones, the skills of the construction, and the mystery surrounds it make the Stonehenge a famous site in England, attracting a large number of tourists each year.

不管最初建造目的是什么，巨石阵规模之大、巨石之美、建造技艺之高超、背景之神秘成为了英格兰的著名景点，每年吸引着大批游客。

④ The tower also proved that France, 100 years after the revolution, took the lead in the field of science and engineering, capable of erecting the tallest edifice.

铁塔还向世人展示了法国大革命之后的一百年，法国在科学、工程学领域处于领先地位，能够树立起世界上最高的建筑。

⑤ It was agreed upon that American people were to build the pedestal, and the French people were responsible for the Statue and its assembly in the United States.

双方达成一致，由美国负责建造基座部分，由法国负责雕像部分，并负责将他在美国组装起来。

4.5.2 建筑文本实训

4.5.2.1　Text A 中国建筑——苏州园林

（1）文本原文。

苏州地处长江三角洲，地理位置优越，气候湿润，交通便利，是中国著名的历史文化名城，素以众多精雅的园林名闻天下。大部分园林都是贵族和富商私人建造的，有的可以追溯到公元前6世纪。旧时官宦名绅晚年多到苏州择地造园、颐养天年。

明清时期，苏州封建经济文化发展达到鼎盛阶段，造园艺术也趋于成熟，出现了一批园林艺术家，使造园活动达到高潮。最盛时期，苏州的私家园林和庭院达到280余处，至今保存完好并开放的有建于宋代的沧浪亭、网师园，元代的狮子林，明代的拙政园、艺圃，清代的留园、精园、怡园、曲园、听枫园等。其中，拙政园、留园、网师园、环秀山庄因其精美卓绝的造园艺术和个性鲜明的艺术特点于1997年底被联合国教科文组织列为"世界文化遗产"。

苏州园林是一座"充满自然意趣的城市山林"。一般来说，园林由两部分组成——住宅区和花园。园林中增加亭台楼阁、池塘、桥梁、假山、石头和芬芳的花朵，以复制出一个规模小但充满生命的自然环境。园林的艺术布局与中国哲学思想完美结合，展现东方建筑文化。每一个组成部分都有造诣的园丁精心设计，有序安排，展示了他们的创造力。身居闹市的人们一进入园林，便可享受到大自然的"山水林泉之乐"。

苏州园林是文化意蕴深厚的"文人写意山水园"。古代的造园者都有很高的文化修养，能诗善画，造园时多以画为本，以诗为题，通过凿池堆山、栽花种树，创造出具有诗情画意的景观，被称为是"无声的诗，立体的画"。在园林中游赏，犹如在品诗，又如在赏画。园林中的石碑和平行对联或房间的名称等，也表现出园林主人对佛教、道教、儒家思想或哲学的抱负、兴趣和理想或追求。有以清幽的荷香自喻人品（如拙政园"远香堂"），有以清雅的香草自喻性情高洁（如拙政园"香州"），有追慕古人似小船自由漂荡怡然自得（如怡园"画舫斋"），还有表现园主企慕恬淡的四园生活（如网师园"小桃源"）等等。这些诗文题刻与园内的建筑、山水、花木自然和谐地结合在一起，使园林的一山一水、一草一木均能产生深远的意境，徜徉其中，可得到心灵的陶冶和美的享受。

工匠们运用建筑与周围景色相结合的独特技巧，创造出远离世界动荡的宁静天堂花园。参观者常常认为他们已经到达了花园的最后一座建筑，但当他们穿过房间和曲折的走廊时，却发现了另一处和谐的风景。每个房间的门窗都用雕刻装饰。透过雕刻装饰的窗户，一幅郁郁葱葱的绿色水景展现在你眼前，石桥上鸟儿的歌唱也与一条潺潺的小溪应和着。沿着蜿蜒的人行道，游客会看到不同的景色。客厅里摆放着简单但很优雅的红木家具，还有珍贵的字画作品。住宅房间通常用苏州风格的盆景和平行对联装饰。苏州古典园林作为居住和园林艺术的集合体，体现了中国古代长江下游人民的生活习惯和礼仪。

在这个集中的"自然世界",一勺水象征着一个湖,一块拳头大小的石头代表着一座山,人们不必出城便可随着时间和季节的变化"享受山林的美丽和宁静,享受泉水和小溪带来的乐趣。"

拙政园以其布局的山岛、竹坞、松岗、曲水之趣,被胜誉为"天下园林之母"。

拙政园是苏州古典园林中面积最大的古典山水园林,全园占地62亩,分为东、中、西和住宅四个部分。住宅是典型的苏州民居,现布置为园林博物馆展厅。拙政园的布局疏密自然,其特点是以水为主,水面广阔,景色平淡天真、疏朗自然。它以池水为中心,楼阁轩榭建在池的周围,其间有漏窗、回廊相连。园内的山石、古木、绿竹、花卉,构成了一幅幽远宁静的画面,代表了明代园林建筑风格。拙政园形成的湖、池、涧等不同的景区,把风景诗、山水画的意境和自然环境的实境再现于园中,富有诗情画意。淼淼池水以闲适、旷远、雅逸和平静氛围见长,曲岸湾头,来去无尽的流水,蜿蜒曲折、深容藏幽而引人入胜;平桥小径为其脉络,长廊逶迤填虚空,岛屿山石映其左右,使貌若松散的园林建筑各具神韵。整个园林建筑仿佛浮于水面,加上木映花承,在不同境界中产生不同的艺术情趣,如夏日蕉廊,冬日梅影雪月,春日繁花丽日,秋日红蓼芦塘,无不四时宜人,不愧为江南园林的典型代表。

(2) 参考译文。

Gardens in Suzhou

Suzhou is located in the Yangtze River delta with a moist climate and convenient transportation. As the famous historical and cultural city, it is often world – famous for a lot of extremely elegant gardens. Most of the gardens were built for private use by aristocrats and rich businessmen and some date back to the sixth Century BC. Officials and celebrities in the old times inclined to build gardens in Suzhou to enjoy their late years.

In the Ming and Qing dynasties, the feudalistic economy and culture in Suzhou reached its summit, leading garden art to maturity. There appeared a large number of garden artists who brought the upsurge of the construction of gardens. During the prime time, there were 280 private gardens and courtyards in Suzhou, among which the well-preserved and opened to the public nowadays are the Canglang Pavilion and the Wangshi Garden (The Master-of-nets Garden) in the Song Dynasty, Shizilin (The Lion Forest Garden) in the Yuan Dynasty, Zhuozheng Garden (The Humble Administrator's Garden), Yipu (The Garden of Cultivation) in the Ming Dynasty, Liuyuan, Jingyuan, Yiyuan, Ouyuan (The Couple's Garden of Retreat), Tingfengyuan in the Qing Dynasty, etc. Zhuozheng Garden, Liuyuan (Lingering Garden), Wangshi Garden and Huanxiu Villa (The Mountain Villa with Embracing Beauty) are incorporated into the "World Cultural Heritage" in the late 1997 by UNESCO for their elaborate art and explicit characteristics.

The gardens in Suzhou are "beautiful urban mountains and forests both with natural beauty and profound cultural implications". In general, the gardens are comprised of two parts — a residential section and a garden. In order to replicate a natural environment on a miniature scale but to be full of life, various kinds of pavilions, ponds, bridges, rockeries, stones and fragrant flowers are added. The artistic layout is combined perfectly with Chinese philosophy and ideology to exhibit an architectural culture of the Orient. Each component in the garden is designed delicately and arranged orderly by the accomplished gardeners, showing their great creativity. When people living in downtown enter the gardens, they would enjoy the beauty of the mountains, lakes, forests and springs in nature.

Suzhou gardens are "freehand landscape gardens of literati" with profound cultural implications. Ancient gardeners had a high cultural accomplishment. They were good at poetry and painting. When gardening, they mostly took painting as the basis and poetry as the theme. They created a poetic landscape by digging pools, piling mountains and planting flowers and trees. Every part was called "silent poetry, three-dimensional painting". Visiting gardens is like appreciating poetry and painting. The names of stone tablets and parallel couplets or rooms in the gardens also show the aspirations, interests and ideals or pursuits of the landscapers towards Buddhism, Taoism, Confucianism or philosophy, such as "Yuanxiangtang" which means the fragrance of the lotus symbolizing the noble human character, "Xiangzhou", which means the elegant sweet grass symbolizing nobility and "Huafangzhai", which describes the crave for the free lives of the ancient people drifting like a boat, and "Zhenyi", "Xiao taoyuan" in Wangshiyuan, which reflect the longing of the garden owner for an idle life in the countryside, etc. These poems and inscriptions are naturally and harmoniously combined with the architecture, landscape, flowers and trees in the garden, so that the landscape of hills, waters, grass and trees can produce a profound artistic conception. Wandering among them, people can get the edification of the soul and the enjoyment of beauty.

Craftsmen use unique techniques to combine the architecture with surrounding scenery and create a peaceful paradise garden far from the turbulence of the world. Visitors often think that they have reached the last building in the garden, but when they cross the room and the winding corridor, they find another harmonious landscape. The doors and windows of each room are decorated with carvings. Through the carved window, a lush green water view unfolds in front of you, and the singing of birds on the stone bridge join in a rippling brook. Walking on a sidewalk winding along a ridge, visitors will see different scenery. In the living room, visitors will see simple but elegant mahogany furniture and exhibition of precious works of calligraphy and painting. Residential rooms are usually decorated with Suzhou style bonsai and parallel couplets. As the aggregation of dwelling and gardening, classical gardens in Suzhou, embody the living habits and etiquette of the people in the lower reaches of the Yangtze River in ancient China.

In this concentrated "natural world", a spoonful of water symbolizes a lake, and a fist - sized stone symbolizes a mountain. People can enjoy the beauty and tranquility of mountains and forests, springs and streams without going out of the city.

Humble Administrator's Garden (Zhuozheng Garden) is praised as "the mother of the world's gardens" for its layout of mountain islands, bamboo docks, pine hills and winding water.

Humble Administrator's Garden is the largest classical landscape garden in Suzhou classical gardens. It covers an area of 62 Mu and is divided into four parts: east, middle, West and the residence. The residence is a typical residential building in Suzhou. It is now arranged as the exhibition hall of the Garden Museum. The layout of Humble Administrator's Garden is dense and natural. Its scenery is plain and natural characterizing in wide water surface. It centers on the pool water, with pavilions built around the pool, connected by hollow - out windows and corridors. The rocks, ancient trees, green bamboo and beautiful flowers in the garden constitute a tranquil scene, which represents the architectural gardens style of the Ming Dynasty. The different scenic spots such as lakes, pools and streams formed by Humble Administrator's Garden reproduce the artistic conception of landscape poems, paintings and the reality of natural environment in the garden, which is full of poetic and picturesque flavor. The water of the pool features in leisure, vastness, elegance and quiet atmosphere. The meandering Bay head, endless flowing water, winding, deep and secluded valleys, are fascinating. The flat bridges work as the venation, with the long corridor filling the void and the islands and rocks adding to its left and right sides, making the loosely – looking garden buildings charming. The whole landscape architecture seems to float on the surface of the water, coupled with the woods and the flowers, producing different artistic interests in different realm: spring with flowers, summer with banana gallery, autumn with red Polygonum and green reeds, winter with plum shadow and snow moon, all of which are timely and pleasant, which deserves to be a typical representative of Jiangnan gardens.

（3）建筑专业术语及词汇。

① 狮子林 The Lion Forest Garden

② 留园 The lingering Garden

③ 拙政园 The Humble Administrator's Garden

④ 沧浪亭 The Canglang Pavilion

⑤ 网师园 The Master – of – nets Garden

⑥ 耦园 The Couple's Garden of Retreat

⑦ 艺圃 The Garden of Cultivation

⑧ 退思园 The Retreat and Reflection Garden

⑨ 环秀山庄 The Mountain Villa with Embracing Beauty

⑩ 长江三角洲 Yangtze River delta

⑪ 湿润的气候 a moist climate

⑫ 封建性的 feudalistic

⑬ 成熟 n. maturity

⑭ 顶点 reach summit

⑮ 建设 construction of gardens

⑯ 院 courtyard

⑰ 亭 Pavilion（台 gazebo/terraces，楼 storeyed building/building，阁 Tower，堂 Hall，轩 windowed veranda，舫 Boat，榭 waterside/pavilions）

⑱ 世界文化遗产 World Cultural Heritage

⑲ 精致的 elaborate

⑳ 城市的 urban

㉑ 曲径通幽 secluded valleys

㉒ 微型的 miniature

㉓ 规模 scale

㉔ 假山 rockeries

㉕ 布局 layout

㉖ 建筑文化 architectural culture

㉗ 山水花园 landscape gardens

㉘ 挖池 digging pools

㉙ 堆山 piling mountains

㉚ 石碑 stone tablets

㉛ 对联 parallel couplets

㉜ 佛教 Buddhism

㉝ 道教 Taoism

㉞ 儒学 Confucianism

㉟ 悠闲的 idle

㊱ 碑文 inscriptions

㊲ 工匠 craftsmen

㊳ 走廊 corridor

㊴ 雕刻 carvings

㊵ 红木家具 mahogany furniture

㊶ 书画作品 works of calligraphy and painting

㊷ 装饰 decorate

㊸ 盆景 bonsai

㊹ 船坞，码头 dock

㊺ 居住区 residence

㊻ 雕窗 hollow-out windows

㊼ 景点 scenic spots

㊽ 以…为特色 features in

㊾ 脉络 venation

㊿ 空隙 void

4.5.2.2　Text B 外国建筑——Palace of Westminster

（1）文本原文。

议会大厦

威斯敏斯特宫（Palace of Westminster），又称议会大厦（Houses of Parliament）是英国议会（包括上议院和下议院）的所在地。威斯敏斯特宫是哥特复兴式建筑的代表作之一，1987 年被列为世界文化遗产。该建筑包括约 1100 个独立房间、100 座楼梯和 4.8 公里长的走廊。尽管今天的宫殿基本上由 19 世纪重修而来，但依然保留了初建时的许多历史遗迹，如威斯敏斯特厅（可追溯至 1097 年），今天用作重大的公共庆典仪式，如国葬前的陈列等。

威斯敏斯特宫位于英国伦敦的中心威斯敏斯特市，它坐落在泰晤士河北岸，接近于以白厅为中心的其他政府建筑物。它的西北角的钟楼就是著名的大本钟（大本钟已于 2012 年 6 月，更名为"伊丽莎白塔"）。

建筑风格

从外表来看，其顶部冠以大量小型的塔楼，而墙体则饰以尖拱窗、优美的浮雕和飞檐以及镶有花边的窗户上的石雕饰品。

威斯敏斯特宫的主轴线上是耸立在威斯敏斯特宫入口之上的维多利亚塔（高 104 米）和大本钟塔（高 98 米）。重量超过 13 吨的大钟得名于一位叫本杰明·霍尔的公共事务大臣。有 4 个直径 9 米的钟盘大钟是在著名的天文学家艾里的领导下建造的。当大钟鸣响报时的时候，钟声通过英国广播公司（BBC）电台响彻四方。

建筑师查尔斯·柏利巴里是一位古典建筑师，但他得到了哥特式建筑师奥古斯都·普金的帮助。威斯敏斯特宫和它的塔楼装饰有众多样式独特的石雕，很多富有创造性和想像力的设计都来源于中世纪的艺术。威斯敏斯特宫的内部找不到单色的天花板和墙壁，到处都是雕花的人行道、华盖、像龛，色彩明快的马赛克拼嵌画，大型的水彩壁画，许多房间里铺有黄色、天蓝色和褐色地板砖。

议会上院天花板完全被出现在徽章中的鸟、动物、花草等形象的浮雕像所覆盖。墙体装有木制墙裙浮雕，墙裙上还有 6 幅水彩壁画。18 位迫使国王签署《英国自由大宪章》的勋爵们的青铜雕像则摆在窗间的像龛内，仿佛是在监视着国王宝座顶的华盖，监视着一排排的裹着鲜红的皮革座椅，监视着上议院议长兼大法官的羊毛口袋。

建筑特点

建筑重建时，查尔斯·巴里爵士的方案运用了垂直哥特风格。该风格曾在 15 世纪和 19 世纪哥德复兴式建筑兴起时风行一时。

建筑外观

石料

威斯敏斯特宫外面的石料最初取自南约克郡小镇阿斯顿的采石场，采用沙色镁质石灰岩。然而污染和石料低品质却导致了石料的剥离和崩坏。虽然早在 1849 年就已发现这些缺陷，然而截至 19 世纪结束，都未有任何举措。直到 1910 年，才确定一些石料需要得到更换。

1928 年，拉特兰产蜜色石灰岩被视为崩坏石料替代品。更换工程始于 1930 年，完工于 1950 年，期间由于受二战影响而一度停工。而到了 1960 年，长期的污染再次造成了损害。1981 年，一项针对外观和高塔的保护修复项目开始实施，1994 年完工。内部庭院的修复将会持续到 2010 年左右。

维多利亚塔和大本钟

查尔斯·巴里爵士的威斯敏斯特宫设计包括了数座塔楼。最高的当属西南广场的维多利亚塔，达到了 98.5 米。其顶部有金属旗杆，王室列席时悬挂皇家旗或在平时悬挂英国国旗。塔基部分是皇家专属通道，用于保障国会开幕大典或是其他官方庆典时期皇室成员的进出。

穿过宫殿中部，很快就能抵达中央厅，中部有一座高 91.4 米（300 英尺）的八角形塔楼，也是威斯敏斯特宫三座主要塔楼中最矮的一座。不同于另外两座，中央塔楼具有拥有一座尖顶，被设计成高层进气口。

宫殿东北角就是最为著名的威斯敏斯特宫钟塔，高 96.3 米（316 英尺）。普金为钟楼的绘图是他为巴里所作的最后一项工作。钟楼顶部的钟房是一座巨大的矩形四面时钟，同样也由普金设计。钟楼拥有 5 座时钟，每过一刻都会报时。最有名的一座为大本钟，每过一小时击打一次。它也是英格兰第三重量的钟表，重达 13.8 吨。尽管"大本"原指该钟表本身，今天已经被人们习惯用来称呼整座塔楼。

另外的圣斯蒂芬塔是一座小型塔楼，位于宫殿前端，威斯敏斯特厅和旧官院间，它的基座囊括了进入下院的主要通道——为人们所熟知的"圣斯蒂芬通道"。其他的塔楼包括有北端的议长塔（Speaker's）和南端的大臣塔（Chancellor's）。分别以重修时下院议长和上院大法官命名。

平面设置

威斯特敏斯特宫有若干小型花园环绕其间。位于宫殿南侧河畔的维多利亚塔花园作为公园对公众开放。黑杖侍卫花园（因黑杖侍卫办公室而命名）紧挨公园，作为专用通道使用。大厦前的旧官院（Old Palace Yard）前有混凝土堆砌的安保屏障。克伦威尔绿地、新官院（北部）和议长绿地（宫殿正北）均为内部场地，不对外开放。上议院的另一侧为克林格绿地（College Green，一块小型三角形绿地，通常作为对政要的电视采访场地）。

（2）参考译文。

The Palace of Westminster, also known as the House of Parliament, is the seat of the British Parliament (including the House of Lords and the House of Commons). Westminster Palace is one of the representative works of Gothic Renaissance architecture, which was listed as a world cultural heritage in 1987. The building consists of about 1100 separate rooms, 100 stairs and 4.8 kilometers of corridors. Although today's palaces were largely rebuilt in the 19th century, many of the original historical relics, such as the Westminster Hall (dating back to 1097), are still preserved. Today they are used for major public ceremonies, such as pre-funeral displays.

Westminster Palace is located in the city of Westminster, center of London, UK. It is on the North Bank of the Thames River, close to other government buildings centered on Whitehall. The bell tower in its northwest corner is the famous Big Ben Bell (Big Ben Bell was renamed "Elizabeth Tower" in June 2012).

Architectural style

Outwardly, the top of Westminster Palace is crowned with a large number of small towers, while the walls are decorated with pointed arch windows, graceful reliefs and eaves, and stone carvings on laced windows.

The Victoria Tower (104 meters high) and Big Ben Bell Tower (98 meters high) are on the main axis of Westminster Palace, which stand above the entrance of Westminster Palace. The bell, weighing more than 13 tons, was named after Benjamin Hall, the Minister of public affairs. Four large clocks with a diameter of 9 metres were built under the leadership of the famous astronomer Ali. When the bell struck the hour, it rang everywhere through the BBC.

Architect Charles Burley was a classical architect, but he was aided by the Gothic architect Augustus Pugin. Westminster Palace and its towers are decorated with many unique stone carvings, many of which are Many creative and imaginative designs derived from medieval art. There were no monochrome ceilings or walls. Instead, carved sidewalks, gables, niches, bright mosaics, large watercolor murals, and yellow, sky-blue and brown floor tiles are found in many rooms.

The ceiling of the upper house of Parliament is completely covered by relief statues of birds, animals, flowers and other images appearing in the emblem. The walls are embossed with wooden breast wall, and there are six watercolor murals on the breast wall. The bronze statues of the 18 lords who forced the King to sign the Great Charter of Freedom were placed in the niches between the windows, as if they were monitoring the canopy of the King's throne, the rows of leather seats wrapped in bright red, and the wool pockets of the Speaker and Chancellor of the House of Lords.

Sir Charles Barry used the vertical Gothic style when the building was rebuilt. This style was popular in the 15th and 19th centuries when Gothic Renaissance architecture came into being.

Exterior

① Stone

The stone outside Westminster Palace was originally taken from a quarry in Aston, a small town in South Yorkshire, using sandy magnesia limestone. However, pollution and low quality of stone lead to the peeling and collapse of stone. Although these defects were discovered as early as 1849, no action had been taken by the end of the 19th century. It was not until the 1910s that it was determined that some of the stones needed to be replaced.

In 1928, the honey-colored limestone produced in Lateran was regarded as a substitute for the collapsed stone. The replacement project began in the 1930s and was completed in the 1950s, during which it was suspended due to the impact of World War II. In the 1960s, long-term pollution caused damage again. In 1981, a project for the protection and restoration of the exterior and tower was launched and completed in 1994. The restoration of the inner courtyard will last until around 2010.

② Tower of Victoria and Big Ben

Sir Charles Barry's Westminster Palace design includes several towers. The highest is the Victoria Tower in the Southwest Square, reaching 98.5 meters. It has a metal flagpole on its top, and the royal flag is hung when the royal family is in attendance or the British flag is hung on all other days. The tower is part of the Royal exclusive corridor, which is used to ensure access to members of the royal family during the opening ceremonies of Congress or other official celebrations.

Crossing the middle of the palace, in the Central Hall, there is an octagonal 91.4 meters (300 feet) high tower, the lowest of the three main towers of Westminster Palace. Unlike the other two, the central tower has a spire and is designed as a high-rise air intake.

The northeastern corner of the palace is the most famous bell tower of Westminster Palace, 96.3 meters (316 feet) high. Putin's drawing of the bell tower was his last job for Barry. The bell house on the top of the bell tower is a huge rectangular four-sided clock, which is also designed by Putin. The Bell Tower has five clocks, which tell the time every 15 minutes. The most famous one is Big Ben, which strikes every hour. It is also the third heaviest clock in England, weighing 13.8 tons. Although "Big Ben" originally refers to the clock itself, today it has been used to call the whole tower.

In addition, St. Stephen's Tower is a small tower, located in the front of the palace, between the Westminster Hall and the old palace courtyard. Its base includes the main access to the lower house, known as the "St. Stephen's Passage". Other towers include Speaker's in the north and Chancellor's in the south. They are named after the Speaker of the House of Commons and the Lord Chancellor of the House of Justice at the time of the renovation.

③ Plan

Westminster Palace is surrounded by several small gardens. The Vitoria Tower Garden on the river south of the palace is open to the public as a park. The Black Staff Guard Garden (named after the Black Staff Guard's Office) is adjacent to the park and is used as a dedicated corridor. The Old Palace Yard in front of the building has a concrete security barrier. Cromwell Greenbelt, New Palace (North) and Speaker's Greenbelt (North of Palace) are all internal venues and are not open to the outside world. On the other side of the House of Lords is College Green, a small triangular green space usually used for television interviews with politicians.

（3）Terms and Vocabularies。

① the Palace of Westminster 威斯敏斯特宫

② the House of Parliament 议会大厦

③ the House of Lords 上议院

④ the House of Commons 下议院

⑤ Gothic a. 哥特式的

⑥ Renaissance 文艺复兴

⑦ world cultural heritage n. 世界文化遗产

⑧ corridor n. 走廊，长廊

⑨ the Thames 泰晤士河

⑩ the Big Ben Bell 大本钟

⑪ pointed arch n. 尖拱

⑫ relief n. 浮雕

⑬ stone carving n. 石雕

⑭ axis n. 轴线

⑮ diameter n. 直径

⑯ medieval a. 中世纪的

⑰ monochrome a. 单色的

⑱ ceiling n. 顶棚，天花板

⑲ gable n. 山墙

⑳ niche n. 壁龛

㉑ mosaic n. 马赛克

㉒ watercolor a. 水彩的

㉓ mural n. 壁画

㉔ tile n. 瓦片

㉕ emblem n. 象征

㉖ breast wall n. 护壁

㉗ canopy n. 华盖

㉘ quarry n. 采石场

㉙ magnesia n. 氧化镁

㉚ limestone n. 石灰岩

㉛ substitute n. 替代品

㉜ flagpole n. 旗杆

㉝ octagonal a. 八角形的

㉞ spire n. 尖顶

㉟ rectangular a. 矩形的

㊱ adjacent a. 临近的

㊲ venue n. 场地

㊳ triangular a. 三角形的

㊴ historical relics 历史遗迹

㊵ pre – funeral display 葬前陈列

㊶ laced window 镶有花边的窗户

㊷ classical architect 古典建筑师

㊸ medieval art 中世纪艺术

㊹ be embossed with 饰以浮雕

㊺ bronze statue 铜像

㊻ vertical Gothic 垂直哥特式风格

㊼ exterior 外观

㊽ peeling and collapse 剥离和崩坏

㊾ inner courtyard 内部庭院

㊿ exclusive corridor 专属通道

4.5.3 翻译练习

4.5.3.1 句子翻译

1. 汉译英

① 苏州地处长江三角洲，地理位置优越，气候湿润，交通便利，是中国著名的历史文化名城，素以众多精雅的园林名闻天下。

② 旧时官宦名绅晚年多到苏州择地造园、颐养天年。

③ 明清时期，苏州封建经济文化发展达到鼎盛阶段，造园艺术也趋于成熟，出现了一批园林艺术家，使造园活动达到高潮。

④ 苏州园林是"充满自然意趣的城市山林"。

⑤ 园林中增加亭台楼阁、池塘、桥梁、假山、石头和芬芳的花朵，以复制出一个规模小但充满生命的自然环境。

⑥ 园林的艺术布局与中国哲学思想完美结合，展现东方建筑文化。

⑦ 每一个组成部分都由有造诣的园丁精心设计，有序安排，展示了他们的创造力。

⑧ 古代的造园者都能诗善画，造园时多以画为本，以诗为题，通过凿池堆山、栽花种树，创造出具有诗情画意的景观。

⑨ 这些诗文题刻与园内的建筑、山水、花木自然和谐地结合在一起，使园林的一山一水、一草一木均能产生深远的意境。

⑩ 徜徉其中，可得到心灵的陶冶和美的享受。

⑪ 工匠们运用建筑与周围景色相结合的独特技巧，创造出远离世界动荡的宁静天堂花园。

　　⑫ 参观者常常认为他们已经到达了花园的最后一座建筑，但当他们穿过房间和曲折的走廊时，却发现了另一处和谐的风景。

　　⑬ 苏州古典园林作为居住和园林艺术的集合体，体现了中国古代长江下游人民的生活习惯和礼仪。

　　⑭ 人们不必出城便可随着时间和季节的变化"享受山林的美丽和宁静，享受泉水和小溪带来的乐趣。"

　　⑮ 拙政园以其布局的山岛、竹坞、松岗、曲水之趣，被胜誉为"天下园林之母"。

　　⑯ 拙政园的布局疏密自然，其特点是以水为主，水面广阔，景色平淡天真、疏朗自然。

　　⑰ 园内的山石、古木、绿竹、花卉，构成了一幅幽远宁静的画面，代表了明代园林建筑风格。

　　⑱ 拙政园形成的湖、池、涧等不同的景区，把风景诗、山水画的意境和自然环境的实境再现于园中，富有诗情画意。

　　⑲ 森森池水以闲适、旷远、雅逸和平静氛围见长。

　　⑳ 整个园林建筑仿佛浮于水面，加上木映花承，在不同境界中产生不同的艺术情趣。

2. 英译汉

① The Palace of Westminster, also known as the House of Parliament, is the seat of the British Parliament (including the House of Lords and the House of Commons).

② Westminster Palace is one of the representative works of Gothic Renaissance architecture, which was listed as a world cultural heritage in 1987.

③ Although today's palaces were largely rebuilt in the 19th century, many of the original historical relics, such as the Westminster Hall (dating back to 1097), are still preserved.

④ It is on the North Bank of the Thames River, close to other government buildings centered on Whitehall.

⑤ Outwardly, the top of Westminster Palace is crowned with a large number of small towers, while the walls are decorated with pointed arch windows, graceful reliefs and eaves, and stone carvings on laced windows.

⑥ The Victoria Tower (104 meters high) and Big Ben Bell Tower (98 meters high) are on the main axis of Westminster Palace, which stand above the entrance of Westminster Palace.

⑦ The bell, weighing more than 13 tons, was named after Benjamin Hall, the Minister of public affairs. Four large clocks with a diameter of 9 metres were built under the leadership of the famous astronomer Ali.

⑧ Westminster Palace and its towers are decorated with many unique stone carvings, many of which are many creative and imaginative designs derived from medieval art.

⑨ The ceiling of the upper house of Parliament is completely covered by relief statues of birds, animals, flowers and other images appearing in the emblem.

⑩ The bronze statues of the 18 lords who forced the King to sign the Great Charter of Freedom were placed in the niches between the windows, as if they were monitoring the canopy of the King's throne, the rows of leather seats wrapped in bright red, and the wool pockets of the Speaker and Chancellor of the House of Lords.

⑪ This style was popular in the 15th and 19th centuries when Gothic Renaissance architecture came into being.

⑫ However, pollution and low quality of stone lead to the peeling and collapse of stone.

⑬ It was not until the 1910s that it was determined that some of the stones needed to be replaced.

⑭ In 1928, the honey-colored limestone produced in Lateran was regarded as a substitute for the collapsed stone.

⑮ The replacement project began in the 1930s and was completed in the 1950s, during which it was suspended due to the impact of World War II.

⑯ In 1981, a project for the protection and restoration of the exterior and tower was launched and completed in 1994.

⑰ It has a metal flagpole on its top, and the royal flag is hung when the royal family is in attendance or the British flag is hung on all other days.

⑱ The tower is part of the Royal exclusive corridor, which is used to ensure access to members of the royal family during the opening ceremonies of Congress or other official celebrations.

⑲ The most famous one is Big Ben, which strikes every hour. It is also the third heaviest clock in England, weighing 13.8 tons.

⑳ The Black Staff Guard Garden (named after the Black Staff Guard's Office) is adjacent to the park and is used as a dedicated corridor.

4.5.3.2 段落翻译

1. 汉译英

① 苏州地处水乡，湖沟塘堰星罗棋布，极利因水就势造园，附近又盛产太湖石，适合堆砌玲珑精巧的假山，可谓得天独厚；苏州地区历代百业兴旺，官富民殷，完全有条件追求高质量的居住环境；加之苏州民风历来崇尚艺术，追求完美，千古传承，长盛不衰，无论是乡野民居，还是官衙贾第，其设计建造皆一丝不苟，独运匠心。这些基本因素大大促进了苏州园林的发展。

② 与欧洲和阿拉伯风格一样，古老的中国建筑风格是全球建筑体系的一个基本要素。在其漫长的发展过程中，它慢慢发展成为一个

以木结构为特点的设计,融合了石刻、夯土建筑、箱式弧形结构以及其他几种策略。艰苦奋斗的中国人创造了许多建筑奇迹,如长城、紫禁城以及秦始皇陵。中国古代建筑的背景久远,可以追溯到商朝(公元前16世纪~公元前771年)。它有自己的框架和格式概念。随着漫长的发展,一些奇迹实际上是由勤劳又聪明的奋斗者创造出来的。古老的建筑奇观数不胜数,从长城、白马寺、莫高窟到颐和园,还有紫禁城。

③ 苏州古典园林的重要特色之一,在于它不仅是历史文化的产物,同时也是中国传统思想文化的载体。表现在园林厅堂的命名、匾额、楹联、书条石、雕刻、装饰,以及花木寓意、叠石寄情等,不仅是点缀园林的精美艺术品,同时储存了大量的历史、文化、思想和科学信息,其物质内容和精神内容都极其深广。其中有反映和传播儒、释、道等各家哲学观念、思想流派的;有宣扬人生哲理,陶冶高尚情操的;还有借助古典诗词文学,对园景进行点缀、生发、渲染,使人于栖息游赏中,化景物为情思,产生意境美,获得精神满足。而园中汇集保存完好的中国历代书法名家手迹,又是珍贵的艺术品,具有极高的文物价值。另外,苏州古典园林作为宅园合一的第宅园林,其建筑规制又反映了中国古代江南民间起居休憩的生活方式和礼仪习俗,是了解和研究古代中国江南民俗的实物资料。

2. 英译汉

① The "Dancing House" is set on a property of great historical significance. Its site was the location of a house destroyed by the U. S. bombing of Prague in 1945. The Dutch insurance company Nationale – Nederlanden (ING Bank since 1991) agreed to sponsor the building of a house on site. The "superbank" chose Milunić as the lead designer and asked him to partner with another world – renowned architect to approach the process. The French architect Jean Nouvel turned down the idea because of the small square footage, but the well – known Canadian – American architect

Frank Gehry accepted the invitation. Because of the bank's excellent financial state at the time, it was able to offer almost unlimited funding for the project. From their first meeting in 1992 in Geneva, Gehry and Milunić began to elaborate Milunić's original idea of a building consisting of two parts, static and dynamic ("yin and yang"), which were to symbolize the transition of Czechoslovakia from a communist regime to a parliamentary democracy.

② The style is known as deconstructivist ("new – baroque" to the designers) architecture due to its unusual shape. The "dancing" shape is supported by 99 concrete panels, each a different shape and dimension. On the top of the building is a large twisted structure of metal nicknamed 'Mary'. In the middle of a square of buildings from the eighteenth and nineteenth century, the Dancing House has two main parts. The first is a glass tower that narrows at half its height and is supported by curved pillars; the second runs parallel to the river and is characterized by undulating mouldings and unaligned windows. The famous dancers Fred Astaire and Ginger Rogers are represented in the structure. A tower made of rock is used to represent Fred. This tower also includes a metal head. A tower made of glass is used to represent Ginger. This design was driven mainly by aesthetic considerations: aligned windows would make evident that the building has two more floors, although it is the same height as the two adjacent nineteenth century buildings. The windows have protruding frames, such as those of paintings, as the designer intended for them to have a three – dimensional effect. The winding mouldings on the façade also serve to confuse perspective effect and diminish contrast with the surrounding buildings.

③ The British Architect, Eva Jiřičná, designed most of the interior. The building has 9 floors. The building also consists of two floors underground. The layout of each of the floors varies due to the asymmetric shape of the building, causing the rooms inside to also be asymmetric. The commercial areas of the building are in the lobby and the first floor. The six floors above are used primarily as offices. The ninth floor housed a restaurant. Since the building takes a slim shape, and the building is split into two parts vertically, the office space is limited. To make the most use of the space, architect Eva Jiřičná used design elements common in ships and incorporated small hallways into the interior of the building. The total interior of the building is 3796 sqm. In 2016, over a course of 5 months, 2 floors of the building were renovated into a 21 – room hotel by Luxury Suites S. R. O. The hotel also has apartments available in each of the tower names after Fred and Ginger. The Ginger & Fred Restaurant now operates on the seventh floor. There is now a glass bar on the eighth floor and an art gallery is also in the building now. The general shape of the building is now featured on a gold 2000 Czech koruna coin issued by the Czech National Bank. The coin completes a series called "Ten Centuries of Architecture". The Dancing House won Time Magazine's design contest in 1997.

4.5.4 练习答案

4.5.4.1 句子翻译

1. 汉译英答案

① Suzhou is located in the Yangtze River delta with a moist climate and convenient transportation. As the famous historical and cultural city, it is often world – famous for a lot of extremely elegant gardens.

② Officials and celebrities in the old times inclined to build gardens in Suzhou to enjoy their late years.

③ In the Ming and Qing dynasties, the feudalistic economy and culture in Suzhou reached its summit, leading garden art to maturity. There appeared a large number of garden artists who brought the upsurge of the construction of gardens.

④ The gardens in Suzhou are "beautiful urban mountains and forests both with natural beauty and profound cultural implications".

⑤ In order to replicate a natural environment on a miniature scale but to be full of life, various kinds of pavilions, ponds, bridges, rockeries, stones and fragrant flowers are added.

⑥ The artistic layout is combined perfectly with Chinese philosophy and ideology to exhibit an architectural culture of the Orient.

⑦ Each component in the garden is designed delicately and arranged orderly by the accomplished gardeners, showing their great creativity.

⑧ Ancient gardeners were good at poetry and painting. When gardening, they mostly took painting as the basis and poetry as the theme. They created a poetic landscape by digging pools, piling mountains and planting flowers and trees.

⑨ These poems and inscriptions are naturally and harmoniously combined with the architecture, landscape, flowers and trees in the garden, so that the landscape of hills, waters, grass and trees can produce a profound artistic conception.

⑩ Wandering among them, people can get the edification of the soul and the enjoyment of beauty.

⑪ Craftsmen use unique techniques to combine the architecture with surrounding scenery and create a peaceful paradise garden far from the turbulence of the world.

⑫ Visitors often think that they have reached the last building in the garden, but when they cross the room and the winding corridor, they find another harmonious landscape.

⑬ As the aggregation of dwelling and gardening, classical gardens in Suzhou, embody the living habits and etiquette of the people in the lower reaches of the Yangtze River in ancient China.

⑭ People can enjoy the beauty and tranquility of mountains and forests, springs and streams without going out of the city.

⑮ Humble Administrator's Garden (Zhuozheng Garden) is praised as "the mother of the world's gardens" for its layout of mountain islands, bamboo docks, pine hills and winding water.

⑯ The layout of Humble Administrator's Garden is dense and natural. Its scenery is plain and natural characterizing in wide water surface.

⑰ The rocks, ancient trees, green bamboo and beautiful flowers in the garden constitute a tranquil scene, which represents the architectural gardens style of the Ming Dynasty.

⑱ The different scenic spots such as lakes, pools and streams formed by Humble Administrator's Garden reproduce the artistic conception of landscape poems, paintings and the reality of natural environment in the garden, which is full of poetic and picturesque flavor.

⑲ The water of the pool features in leisure, vastness, elegance and quiet atmosphere.

⑳ The whole landscape architecture seems to float on the surface of the water, coupled with the woods and the flowers, producing different artistic interests in different realm.

2. 英译汉答案

① 威斯敏斯特宫，又称议会大厦，是英国议会（包括上议院和下议院）的所在地。

② 威斯敏斯特宫是哥特复兴式建筑的代表作之一，1987年被列为世界文化遗产。

③ 尽管今天的宫殿基本上由19世纪重修而来，但依然保留了初建时的许多历史遗迹，如威斯敏斯特厅（可追溯至1097年）。

④ 威斯敏斯特宫位于英国伦敦的中心威斯敏斯特市，它坐落在

泰晤士河北岸，接近于以白厅为中心的其他政府建筑物。

⑤ 从外表来看，其顶部冠以大量小型的塔楼，而墙体则饰以尖拱窗、优美的浮雕和飞檐以及镶有花边的窗户上的石雕饰品。

⑥ 威斯敏斯特宫的主轴线上是耸立在威斯敏斯特宫入口之上的维多利亚塔（高104米）和大本钟塔（高98米）。

⑦ 重量超过13吨的大钟得名于一位叫本杰明·霍尔的公共事务大臣。有4个直径9米的钟盘大钟是在著名的天文学家艾里的领导下建造的。

⑧ 威斯敏斯特宫和它的塔楼装饰有众多样式独特的石雕，很多富有创造性和想像力的设计都来源于中世纪的艺术。

⑨ 议会上院天花板完全被出现在徽章中的鸟、动物、花草等形象的浮雕像所覆盖。

⑩ 18位迫使国王签署《英国自由大宪章》的勋爵们的青铜雕像则摆在窗间的像龛内，仿佛是在监视着国王宝座顶的华盖，监视着一排排的裹着鲜红的皮革座椅，监视着上议院议长兼大法官的羊毛口袋。

⑪ 该风格曾在15世纪和19世纪哥德复兴式建筑兴起时风行一时。

⑫ 然而污染和石料低品质却导致了石料的剥离和崩坏。

⑬ 直到1910年，才确定一些石料需要得到更换。

⑭ 1928年，拉特兰产蜜色石灰岩被视为崩坏石料替代品。

⑮ 更换工程始于1930年，完工于1950年，期间由于受二战影响而一度停工。

⑯ 1981年，一项针对外观和高塔的保护修复项目开始实施，1994年完工。

⑰ 其顶部有金属旗杆，王室列席时悬挂皇家旗或在平时悬挂英国国旗。

⑱ 塔基部分是皇家专属通道，用于保障国会开幕大典或是其他

官方庆典时期皇室成员的进出。

⑲ 最有名的一座为大本钟,每过一小时击打一次。它也是英格兰第三重量的钟表,重达 13.8 吨。

⑳ 黑杖侍卫花园(因黑杖侍卫办公室而命名)紧挨公园,作为专用通道使用。

4.5.4.2 段落翻译

1. 汉译英答案

① Suzhou is situated in a watery country, where lakes, gullies, ponds and weirs are scattered all over the country. And also with plenty of Taihu Stones nearby, it has become a unique place to build exquisite rockeries. Suzhou has flourished in all dynasties, with rich officials and people, so is fully equipped to pursue a high – quality living environment. In addition, people in Suzhou have always advocated art, pursued perfection, inheritance and prosperity. Country dwellings, the buildings of government and residence of rich businessmen, all are meticulously designed and constructed with ingenuity. These basic factors have greatly promoted the development of Suzhou gardens.

② Along with European and also Arabian style, old Chinese architecture style is an essential element of the globe building system. Throughout its lengthy advancement, it slowly developed right into a design which featured in timberwork integrating rock sculpting, rammed – earth building, container arc structures as well as several various other strategies. Laborious Chinese produced numerous building wonders such as the Great Wall, Forbidden City as well as the Mausoleum of the First Qin Emperor. Ancient Chinese architecture has a lengthy background which could be mapped back to the Shang Empire (16th century BC – 771 BC). It has its very own concepts of framework as well as format. With the lengthy advancement, several marvels have actually been developed by laborious and smart Chinese people. Old building wonders are countless varying from Great Wall, White Steed Holy Place, Mogao Caves to Summer Palace as well as Forbidden City.

③ One of the important characteristics of Suzhou classical gardens is that it is not only the product of history and culture, but also the carrier of Chinese traditional ideology and values. The naming, plaque, couplet, calligraphy stone, sculpture, decoration, flower and trees, stones, etc. are not only the beautiful works of art embellishing the garden, but also the storage of a large number of historical, cultural, ideological and scientific information. Their material and spiritual contents are extremely profound. Some of them reflect and disseminate the philosophical concepts and schools of thought of Confucianism, Buddhism and Taoism, while others advocate the philosophy of life and cultivate noble sentiments. And the classical poetry and literature used in gardens have embellished, innovated and rendered the garden scenery, which produced artistic beauty made people interpret the scenery with different feelings and thus obtain spiritual satisfaction. The garden has a collection of well-preserved handwriting of famous Chinese calligraphers of past dynasties and has a very high value of cultural relics. In addition, Suzhou classical gardens, as the residential gardens, reflect the life style and etiquette customs of the folk living and recreation in ancient China, and are the physical materials for understanding and studying the folk customs of ancient China in the south of the Yangtze River.

2. 英译汉答案

①"跳舞房子"坐落在一个具有重大历史意义的地方，它的所在地是1945年美国轰炸布拉格时摧毁的一座房子。荷兰保险公司Nationale Nederlanden（1991年成立荷兰银行）同意在现场赞助建造一

座房屋。"超级银行"选择了米奴尼（Miluni）作为首席设计师，并要求他与另一位世界著名的建筑师合作来进行设计。法国建筑师让·努维尔（Jean Nouvel）因为占地面积小而拒绝了这个想法，但著名的加拿大裔美国建筑师弗兰克·盖里（Frank Gehry）接受了邀请。由于当时银行财务状况良好，为该项目提供了几乎无限的资金支持。从1992年在日内瓦召开的第一次会议开始，盖里和米奴尼就开始关于一座由静态和动态两部分组成的建筑（"阴阳"）的设计构想，使建筑象征着捷克斯洛伐克从共产主义政权到议会民主的过渡。

②"跳舞房子"的风格由于其不寻常的形状被称为解构主义（"新巴洛克"的设计师）建筑。"跳舞房子"的形状由99块混凝土板支撑，每一块都有不同的形状和尺寸。建筑物的顶部是一个巨大的金属扭曲结构，绰号"玛丽"。"跳舞房子"坐落于十八世纪和十九世纪的一个建筑广场中间，有两个主要部分：第一个是一个玻璃塔，它在一半的高度变窄，由弯曲的柱子支撑；第二个塔与河流平行，其特点是造型起伏，窗户没有对齐。著名的舞蹈家弗雷德·阿斯泰尔和金吉尔·罗杰斯都出现在这座建筑中。用石头做成的塔代表弗雷德，这座塔包括一个金属塔头，用玻璃制成的塔代表金吉尔。"跳舞房子"的设计主要是出于美学考虑：尽管与相邻的两座十九世纪建筑高度相同，但对齐的窗户可以明显看出该建筑还有两层楼。设计师希望它们具有三维效果，窗户有突出的框架，如绘画框架。立面上的缠绕造型也会混淆透视效果，降低与周围建筑的对比度。

③ 英国建筑师伊娃·吉恩（Eva Jiin）设计了大部分室内设计。整座大楼有9层楼高，还包括地下两层。由于建筑物的形状不对称，每层楼的布局都不尽相同，导致室内的房间也不对称。大楼的商业区在大厅和一楼，上面六层主要用作办公空间，九楼有一家餐馆。大楼呈细长形状，垂直分为两部分，办公空间有限。为了充分利用空间，建筑师伊娃·吉恩使用了船上常见的设计元素，并将小走廊融入建筑内部，使该建筑的内部总面积达3796平方米。2016年，经过5个月

的时间，这座大楼的两层楼被重新装修成了一家拥有 21 间客房的酒店。七楼是金吉尔和弗雷德餐厅，八楼有一个玻璃酒吧。这栋楼里还有一个美术馆。捷克国家银行发行的一枚价值 2000 捷克克朗的金币上现刻着该建筑的大致形状，这枚硬币包含了一个名为"十世纪建筑"的系列作品。舞厅也被《建筑杂志》评为 20 世纪 90 年代最重要的 5 座建筑之一。

简明建筑英语
翻译教程
Chapter 5

第五部分　建筑文本翻译拓展训练（一）英译汉
——Gothic, Rococo and Mediterranean Architecture

第五部分　建筑文本翻译拓展训练（一）英译汉

5.1　文本原文

1　Gothic Architecture

(1) A Brief Introduction。

Gothic architecture is a style of architecture that flourished during the high and late medieval period. It evolved from Romanesque architecture and was succeeded by Renaissance architecture.

Originating in 12th century France and lasting into the 16th century, Gothic architecture was known during the period as "the French Style", with the term Gothic first appearing during the latter part of the Renaissance. Its characteristic features include the pointed arch, the ribbed vault and the flying buttress.

Gothic architecture is most familiar as the architecture of many of the great cathedrals, abbeys and churches of Europe. It is also the architecture of many castles, palaces, town halls, guild halls, universities and to a less prominent extent, private dwellings. It is in the great churches and cathedrals and in a number of civic buildings that the Gothic style was expressed most powerfully, its characteristics lending themselves to appeal to the emotions. A great number of ecclesiastical buildings remain from this period, of which even the smallest are often structures of architectural distinction while many of the larger churches are considered priceless works of art and are listed with UNESCO as World Heritage Sites. For this reason a study of Gothic architecture is largely a study of cathedrals and churches.

(2) The History of Gothic Architecture。

Gothic architecture grew out of the previous architectural genre, Romanesque. For the most part, there was not a clean break, as there was later to be in Renaissance Florence with the revival of the classical style by Brunelleschi in the early 15th century.

Romanesque tradition

Romanesque architecture, or Norman architecture as it is generally termed in England because of its association with the Norman invasion, had already established the basic architectural forms and units that were to remain in slow evolution throughout the Medieval period. The basic structure of the cathedral church, the parish church, the monastery, the castle, the palace, the great hall and the gatehouse were all established. Ribbed vaults, buttresses, clustered columns, ambulatories, wheel windows, spires and richly carved door tympanums were already features of ecclesiastical architecture.

The widespread introduction of a single feature was to bring about the stylistic change that separates Gothic from Romanesque, and broke the tradition of massive masonry and solid walls penetrated by small openings, replacing it with a style where light appears to triumph over substance. The feature that brought the change is the pointed arch. With its use came the development of many other architectural devices, previously put to the test in scattered buildings and then called into service to meet the structural, aesthetic and ideological needs of the new style. These include the flying buttresses, pinnacles and traceried windows which typify Gothic ecclesiastical architecture.

Possible Eastern influence

While so-called 'pitched' brick vaulting, which could be constructed without centering, may date back in the Ancient Near East to the 2nd millennium BC, the earliest evidence of the pointed masonry arch appears in late Roman and Sassanian architecture, mostly evidenced in early church building in Syria and Mesopotamia, but occasionally also in secular structures like the Karamagara Bridge. After the Muslim conquests of the 7th century, it became gradually a standard feature of Islamic architecture.

According to one theory, increasing military and cultural contacts with the Muslim world, as Norman Conquest of Islamic Sicily in 1090, the Crusades which began in 1096 and the Islamic presence in Spain brought the knowledge of pointed arches to Medieval Europe. According to another theory, it is believed that the pointed arch evolved naturally in Western Europe as a structural solution to a purely technical problem, concurrent with its introduction and early use as a stylistic feature in French and English churches.

Abbot Suger

Abbot Suger, friend and confidante of the French Kings, Louis VI and Louis VII, decided in about 1137, to rebuild the great Church of Saint – Denis, attached to an abbey which was also a royal residence.

Suger began with the West front, reconstructing the original Carolingian façade with its single door. He designed the façade of Saint – Denis to be an echo of the Roman Arch of Constantine with its three – part division and three large portals to ease the problem of congestion. The rose window is the earliest – known example above the West portal in France.

At the completion of the west front in 1140, Abbot Suger moved on to the reconstruction of the eastern end, leaving the Carolingian nave in use. He designed a choir (chancel) that would be suffused with light. To achieve his aims, his masons drew on the several new features which evolved or had been introduced to Romanesque architecture, the pointed arch, the ribbed vault, the ambulatory with radiating chapels, the clustered columns supporting ribs springing in different directions and the flying buttresses which enabled the insertion of large clerestory windows.

The new structure was finished and dedicated on 11 June 1144, in the presence of the King. The Abbey of Saint – Denis thus became the prototype for further building in the royal domain of northern France. It is often cited as the first building in the Gothic style. A hundred years later, the old nave of Saint – Denis was rebuilt in the Gothic style, gaining, in its transepts, two spectacular rose windows. Through the rule of the Angevin dynasty, the style was introduced to England and spread throughout France, the Low Countries, Germany, Spain, northern Italy and Sicily.

(3) The Revival of Gothic Architecture。

A series of Gothic revivals began in mid – 18th century England, spread through 19th – century Europe and continued, largely for ecclesiastical and university structures, into the 20th century.

However, the term Gothic was first used as a term of contempt. Says Vasari, "Then arose new architects who after the manner of their barbarous nations erected buildings in that style which we call Gothic", while Evelyn but expresses the mental attitude of his own time when he writes, "The ancient Greek and Roman architecture answered all the perfections required in a faultless and accomplished building" – but the Goths and Vandals destroyed these and "introduced in their stead a certain fantastical and licentious manner of building: congestions of heavy, dark, melancholy, monkish piles, without any just proportion, use or beauty." For the first time, an attempt was made to destroy an instinctive and, so far as Europe was concerned, an almost universal form of art, and to substitute in its place another built up by artificial rules and premeditated theories; it was necessary, therefore, that the ground should be cleared of a once luxuriant growth that still showed signs of vitality, and to effect this the schools of Vignola, Palladio, and Wren were compelled to throw scorn on the art they were determined to discredit. As ignorant of the true habitat of the style as they were of its nature, the Italians of the Renaissance called it the "maniera Tedesca", and since to them the word Goth implied the perfection of barbarism, it is but natural that they should have applied it to a style they desired to destroy. The style ceased, for the particular type of civilization it expressed had come to an end; but the name remained, and when, early in the nineteenth century, the beginnings of a new epoch brought new apologists, the old title was taken over as the only one available, and since then constant efforts have been made to define it more exactly, to give it a new significance, or to substitute in its place a term more expressive of the idea to be conveyed.

It should be noted that the Gothic Revival did not sweep aside all other styles. Gothic was but one of a number of revival styles popular in the nineteenth century. At the time of his immigration in the early 1890s, G. A. Audsley would have been met by numerous examples of the Romanesque Revival. The style had been pioneered in the United States by Henry H. Richardson (1838 – 1886) who was influenced by historical buildings in southern France. The "Richardsonian" Romanesque was massive in character with roughly cut stone blocks, sometimes in contrasting colors, and monumental round arches. It had virtually become a national style when its popularity plunged with the erection of the Roman Revival pavilions of the Chicago Columbian Exposition in 1893. Judging by his negative comments on Romanesque Revival stonework in the 1890s, G. A. Audsley probably had no interest in this style.

The Gothic Revival proved to be more durable than the Romanesque Revival. The end of the nineteenth century saw major cathedral projects including New York, Washington, and Liverpool, where the actual structural system of Medieval Gothic architecture was revived. The plaster vaults and iron columns of an earlier era were not to be admitted here, and much was made of the fact that these buildings were entirely of stone with no structural steelwork.

Quite surprisingly, the Gothic style found popularity in early twentieth century structures containing a great deal of steel. The vertical emphasis of Gothic architecture was found by American architects to be suited to skyscrapers. The best-known example was Cass Gilbert's (1859 – 1934) Woolworth Building (1913) in New York City. The 880-foot-high facade was composed of terra cotta panels molded with Gothic ornament and supported by the steel frame.

By 1930, the Gothic Revival had largely run its course. Churches and college campuses, the last strongholds of the Medieval Gothic style, likewise sheltered the last of the Gothic Revival. Art Deco and Modernism, both futuristic styles, overtook the traditional.

(4) Gothic Forms。

Plan: the typical plan was a 3 or 5 aisle cross plan which meant that the nave (the central aisle) had one or two aisles on each side. The plan included the transept, choir and sanctuary. During the Gothic period, the ambulatory was developed for circular with small chapels attached to it. The west façade was always the main entrance and usually had one or two monumental towers.

Interior Elevation: Like many Romanesque churches, the interior was divided between the arcade, the triforium and the clerestory. Early Gothic cathedral also had a gallery between the arcade and the triforium, which was later eliminated as the clerestory windows became larger and larger.

Pointed Arch: Unlike the Early Christian, Byzantine and Romanesque semi-circular arch, the pointed Gothic arch was created by the later Medieval architects for many reasons. The semi-circular arch could perfectly vault a square space below, but awkwardly vaulted a rectangular space as the apex of the arches was at different heights. The pointed arch could be built flatter or more pointed without much notice. Thus, the pointed arch more gracefully vaulted rectangular or square spaces. The master builder/architect could now more easily design with both a rectangular or square bay plan. The pointed arch also created less lateral thrust than a round arch so it was easier to support. Overall, the pointed arch looked lighter and more delicate than the more solid and massive circular arches.

Ribbed Vaults: When the pointed arches were combined to make vaults, the vaults were ribbed for increased structural strength. The ribs were nothing more than pointed arches running diagonally (across the nave), transversely (perpendicular with the nave) or longitudinally (parallel with the nave). These structural ribs allowed the infill panels between the ribs to be light-weight stone which reduced the overall weight of the vault. Together with the flying buttresses, the rib vaults also concentrated the forces of the vaults into the exterior piers which allowed the walls to be completely opened up with windows and light. The Gothic church was designed with a skeleton-like framework. Rib vaulting was an organic (natural) metaphor (comparison) alluding to the ribs in the body's skeletal frame.

Flying Buttress: The great lateral forces of the nave vaults were transferred to the exterior piers outside of the building by means of connecting bridges called flying buttresses. The buttresses essentially flew over the aisles to the outer piers. Pinnacles were placed on top of the buttresses to ass weight and additional strength. This new gothic invention allowed the wall area between the buttresses to be filled with windows.

Stone Tracery and Stained Glass: with new structural knowledge, the Gothic builders attempted to create a spiritual force within the building. Translucent walls of stained glass connected by stone tracery (and later lead work) were like walls of lace depicting beautiful patterns or scenes from the bible.

Sculpture: On the exterior, the cathedral was covered with stone sculpture depicting the king and other royals, important clergy and stories from the bible. The intricate stone carving expressed a delicate balance between structure and decoration. The combination of the arts was so complete, that design, structure and art became one.

(5) Gothic Church Architecture。

Gothic church architecture in Medieval England developed from Norman architecture. 'Gothic architecture' is the term used to describe building styles between 1200 to 1500. Such a large time span meant that a number of styles developed within Gothic architecture and it is common to divide these styles into three sections. The building between 1200 to 1300 is usually referred to as Early English; between 1300 to 1400, the style of building is referred to as Decorated and from 1400 to 1500, it is known as Perpendicular. It is common for major church buildings to show examples from all three of these periods.

Gothic cathedrals are characterized by large towers and spires. Whereas Norman architecture can be seen as being 'dumpy' due to their more limited knowledge of building, the Gothic era coincided with a greater knowledge of engineering and this is reflected in the church buildings completed during this era.

Gothic churches and cathedrals were fundamentally different to Norman buildings. The increase in knowledge and skills acquired over the years meant that stone was specifically cut so that it fitted next to other stone blocks with precision. Therefore, the large blocks of stone favored by the Normans, were replaced by shaped stone. Another major change was that the hollow walls used by the Normans were not used by later architects. Walls and pillars were solid and this allowed them to cope with much greater weights. This simple fact allowed churches and especially cathedrals to be much larger than Norman ones. This, along with the money gathering ability of the Church, explains why the cathedrals and churches of the Gothic era were so much larger than previous ones.

Another development that strengthened church buildings was the use of pointed arches. This shape allowed a much greater weight to be carried when compared to a Norman rounded arch. Cathedral roofs were now much larger than Norman roofs. Therefore, they were a lot heavier. To ensure that the walls and pillars could take such a weight, the architects in this era developed what were known as buttresses. These were additions to the main part of the cathedral that allowed the extra weight to be transferred to additional parts of a cathedral than ran alongside the nave and then down into the foundations. The architects simply spread the weight to other points in the building. 'Flying buttresses' allowed the outward pressure of the massive roofs to be resisted.

The concern about the weight of the roof at York Minster was such that the vaults in all but the smallest aisles were made of wood. This decreased the pressure on the pillars, foundations etc but led to future problems concerning fire and death watch beetles. York Minster does have flying buttresses but these were added in the Nineteenth Century.

The ability to cope with greater weights also allowed Gothic architects to use larger windows. The Normans had been limited to using small slit windows. Now cathedrals and churches could have large stained glass windows. The Great East window at York Minster is the size of a tennis court, a size that would have been unthinkable for the Normans.

These new huge buildings cost vast sums of money. Where did the church get this money from? Basically, the bulk of it came from the people of England. Peasants and town dwellers paid numerous taxes to the church –

a tax at baptisms, marriages and deaths; tithes and for centuries people had to work for free on church land. The revenue gained from these assisted the building of cathedrals like those at Lincoln, York, Canterbury and Chichester.

(6) Famous Gothic Cathedrals。

Notre Dame Cathedral, Paris, 1152 – 1240

Notre Dame was an Early Gothic combination of conservative and progressive ideas. The plan was originally 5 – aisled with quadra – partite vaulting over the side aisle bays and sexpartite vaulting over the almost square bays in the nave. The layout was continuous from one end to the other with the side aisles terminating in a double ambulatory and interrupted only by a transept that just barely broke the line of the exterior façade. Nave, choir and transept were distinct spaces, yet harmoniously balanced with smooth transitions. Unlike Durham Cathedral, there was no longer an alternate system of columns and compound piers. At Notre Dame only great columns support cluster of smaller engaged columns which rise to the vaulted ceiling above.

About 1250, Notre Dame was 'modernized' to compete with other new cathedral designs in France. As a result, almost 30 chapels were built in the spaces between the buttresses. Within the transept facades, the great rose windows were built with stone tracery and colorful stained glass. Also at this time, the great spire was added to the crossing. The nave wall was modernized too.

The 4 level elevations (arcade, gallery, triforium and clerestory) had become obsolete (outdated). At this time the gallery (passage above the side aisle) was removed. In an attempt to allow more light into the cathedral, the flying buttresses were expanded so the clerestory windows could also be enlarged. The façade balanced vertical forms against horizontal elements which created a sense of stability. The west façade, call the Royal Gallery, was the first gallery built on a cathedral front. The sculpture represented the French King and his family. On the north and south transept facades, the stone sculpture depicted stories from the Bible. Notre Dame was the first cathedral to be built on a truly monumental scale and became a prototype for future French cathedral.

Chartres Cathedral, Chartres, 1190 – 1220

Completed by 1220, Chartres was considered the first High Gothic cathedral. It was the first cathedral to be planned from the beginning with flying buttresses. An aerial view shows the dramatic buttresses placed around the chevet (the east end of the church), thus supporting the vaults inside with permanent arms – almost like scaffolding left in place.

The plan had a new kind of organization. The square – bay layout of the early Gothic churches was now replaced by a rectangular – bay system in the nave and transept (the aisles still had square bays) that would become typical of the High Gothic style. The High Gothic vault was quadra – partite with only 4 panels which were aligned to create a sequence of continuous volumes. The nave had become a vast continuous hall, surrounded not by heavy walls, but by screens of beautiful colored glass.

Without a separate gallery, the arcade and clerestory were expanded. The clerestory had 2 hall windows (twin lancets) topped by an oculus (flower – shaped window). The interior was filled with light which was dimmed and colored as it came in through the stained glass to create colored light.

The west portal (door) of Chartres was carved in 1150 and told the story of Christ: birth (right), rise to heaven (left) and second coming (center). Scenes from Christ's life were also carved on the capitals in continuous bands from one portal to the next. Other carvings represented the liberal arts, the signs of the zodiac and the labors of the months or seasons. The King and Queen were also depicted in stone. The spires were built at different times in somewhat different styles which help to explain their asymmetry. All three entrances were huge, cavernous (deep) triple doorways.

Even when the church was closed, it was possible for the visitor to receive religious inspiration and education from the carvings and sculpture on the exterior of the church. When the cathedral was open, the stained glass panels also told stories from the Bible. Overall, the cathedral was a not only a place for the religious spirit but also a place religious education.

Amiens Cathedral, Amiens, 1219 – 1269

Amiens was the tallest cathedral built in France, containing the greatest interior space of any High Gothic cathedral. Amiens refined the forms of Chartres. The nave was taller and thinner than ever before and the arcade higher. The thinner of the pier proportions created an impression of colossal scale and vertical lift. It had the most continuous visual connection from floor to vault. The quadra – partite vault of the choir were like a canopy or a tent suspended from giant piers. The light coming in from the clerestory gave vertical lift while blurring the structural outlines. Physical mass was reduced by structural technology and light.

The plan had a rectangular – bay system, 4 – paneled rib vaulting and a buttressing system that permitted abundant light but was more refined and delicate. It was truly a skeletal structure. The walls between the piers were only used to hold the windows and protect the inside from the weather. Above the stone vaults was the steep wooden trussed roof which reached 61m above ground. The steep roofs, which were good at shedding rain and snow, were also good at catching the wind. As a result, two set of flying buttresses were needed to transmit the forces of roof and nave vaults to the external buttress piers. The lower flying buttress transmitted the outward thrust of the nave vaults and the upper flying buttresses carried the wind loads of the tall roofs to the buttresses.

The façade had uneven towers as the shorter tower was built in the 14th century and taller tower in the 15th century. The portals were deeply recessed behind the front plane of the building. The deep piercing of the walls and tower left few undecorated surfaces as the building was covered with colonnades, arches, pinnacles, rose windows and other stonework that almost made the solid structure disappear. The main entrance portals aligned with the aisles and central transept.

2 Rococo Architecture

(1) An Introduction to Rococo。

Rococo, style in architecture, especially in interiors and the decorative arts, which originated in France and was widely used in Europe in the 18th cent. The term may be derived from the French words rocaille and coquille (rock and shell), natural forms prominent in the Italian baroque decorations of interiors and gardens. The first expression of the rococo was the transitional régence style. In contrast with the heavy baroque plasticity and grandiloquence, the rococo was an art of exquisite refinement and linearity. Through their engravings, Juste Aurèle Meissonier and Nicholas Pineau helped spread the style throughout Europe. The Parisian tapestry weavers, cabinetmakers, and bronze workers followed the trend and arranged motifs such as arabesque elements, shells, scrolls, branches of leaves, flowers, and bamboo stems into ingenious and engaging compositions. The fashionable enthusiasm for Chinese art added to the style the whole bizarre vocabulary of chinoiserie motifs. In France, major exponents of the rococo were the painters Watteau, Boucher, and Fragonard and the architects Robert de Cotte, Gilles Marie Oppenord, and later Jacques Ange Gabriel. The rococo vogue spread to Germany and Austria, where François de Cuvilliès was the pioneer. Italian rococo, particularly that of Venice, was brilliantly decorative, exemplified in the paintings of Tiepolo. The furniture of Thomas Chippendale manifested its influence in England. During the 1660s and 1670s, the rococo competed with a more severely classical form of architecture, which triumphed with the accession of Louis XVI.

The Rococo style superseded the Baroque style beginning in France in the late 1720s, especially for interiors, paintings and the decorative arts. Rich Baroque designs were giving way to lighter elements with more curves and natural patterns. The delicacy and playfulness of Rococo designs is often seen as a reaction to the excesses of Louis XIV's regime. The 1730s represented the height of Rococo development in France. The style had spread beyond architecture and furniture to painting and sculpture. Rococo still maintained the Baroque taste for complex forms and intricate patterns. By this point, it had begun to integrate a variety of diverse characteristics, including a taste for Oriental designs and asymmetric compositions.

Fashion in the period 1700 – 1750 in European and European – influenced countries is characterized by a widening, full – skirted silhouette for both men and women following the tall, narrow look of the 1680s and 90s. Wigs remained essential for men of substance, and were often white; natural hair was powdered to achieve the fashionable look.

Distinction was made in this period between full dress worn at Court and for formal occasions, and undress or everyday, daytime clothes. As the decades progressed, fewer and fewer occasions called for full dress which had all but disappeared by the end of the century.

Rococo architecture came about as a reflection of the times. It followed the Baroque style and was known for its feminine curves, intricate designs, and flamboyance. It was a much lighter style of architecture than the dark heavy Baroque style and emphasized by intricate details and very light colors. The style was meant to be a reflection of the times, meaning a time that was frivolous, happy and uneventful.

Rococo architecture became known as the French style and really did not do as well in other countries as it did in France. The Rococo architecture style took its creativity from nature, referring to clouds, flowers, shells, sea, coral, scrolls, spray, etc. Most of the colors that were used in the buildings of the times were pastels or very light colors.

Among some of the most noted buildings of the period and which are still standing are the Hotel de Matignon, and the Hotel d'Evreux, the Place Louis XV designed by Jacques Ange Gabriel which we now know as the Place de la Concorde.

Besides intricate designs and frivolous detail the Rococo architecture also brought many improvements to architecture; sanitation was improved, chimneys were made more efficient and rooms were better organized to offer more privacy.

(2) Rococo's Influence。

The Rococo movement influenced other arts including painting, architecture and sculpture. Its playful nature, delicate strokes and feminine style influenced also greatly furniture, tapestry, interior design and clothing.

The themes became very light-hearted and within the decorative arts most Rococo pieces were very intricate. That is why it worked best with small scaled items, which differed greatly to the large Baroque sculpture and architecture of previous eras. Rococo was best used indoors and was adapted to porcelain figures, frills, metal work, and furniture.

The architectural, interior design and even the clothing of the Rococo style was very common in the Marie Antoinette period. In her palace you will note that the rooms were done entirely in the Rococo style and are considered to be works of art themselves. Rococo furniture is known to be very ornate. Tapestries, mirrors, ornaments and paintings done in the same style were used to complement the Rococo architecture. As the period ended, much of the style was looked upon derogatively and considered to be too frivolous.

(3) Baroque and Rococo。

The French word "baroque" means bizarre or fantastic. Spaces were manipulated like plastic or clay, architectural elements were deliberately distorted, and the laws of nature were often ignored. Baroque architectural elements were deliberately distorted, and the laws of nature were often ignored. Baroque architects had complete artistic freedom. Yet, while Baroque architecture was an expression of unusual creativity and originality, it was also symbolic of extreme decadence and excessive ornamentation. The Rococo was the last phase of the Baroque. During the early 1700's, both the royalty and church leaders in France and Germany demanded more elegant and extravagant architecture. The Rococo was characterized by lightness in color while the Baroque style was much darker and heavier feeling. Rococo building exteriors were also more delicate and refined than the Baroque. Overall, the Rococo was a time of extreme decoration, ornamentation and display of wealth and power.

(4) Examples of Rococo Architecture。

Versailles, France, 1670 – 1720

Around 1670, King Louis XIV (14th) decided to turn a royal hunting lodge, a few miles outside of Paris in Versailles, into a great palace. An army of architects, sculptors, painters and landscape architects were brought together to create the greatest monument of the French Baroque/Rococo. Versailles was a symbol of human power and order over the fragility and disorder of nature.

Of the hundreds of rooms within the palace, the most famous was the Galerie des glaces or Hall of Mirrors which overlooked the park from the second floor and extended along most of the width of the central block. Its enormous size was heightened by hundreds of mirrors, set into the wall opposite the windows which created the illusion of greater width. The mirror, the ultimate source of illusion, was a favorite element of Baroque interior design.

The enormous palace would have appeared unbearable had it not been for its setting in a vast park – like landscape. The park, designed by Andre Le Notre, transformed and entire forest in to a collection of axially arranged gardens, vistas and paths (allies). The gardens created a gradual transition from formal architecture to wild nature. The formal gardens near the palace were tightly designed geometric units but farther away from the palace the design loosened as tree screens or framed views of the distant countryside. No photograph can describe the experience of the design as the park can only be experienced by walking through it. Even though the garden was designed by man, the art of the landscape was purely natural as the garden changed with each hour of the day, the season and the light at that moment.

Bishop's Residence, Wurzburg, 1720 – 1744

Among the most famous German architects of the Rococo era was Johann Balthazar Neumann. It was in the south of Germany that German Baroque and Rococo reached its peak. Theatrical illusionism, the combining of painting, sculpture and architecture, was used to express religious enthusiasm and excessive wealth.

Neumann was a master of bringing light into a building through invisible sources and making that light feel like radiance from heaven. In the Residence (Palace) of the Prince Bishop in Wurzburg, the interiors were the focus of attention as seem in the Main Staircase, the Reception Hall and the Chapel. Neumann raised the idea of the stair from a humble and functional element into something spectacular. The stair climbed up first on a single, central flight and then in the reverse direction, on two wings that doubled back and surrounded the central flight. The ascent was into an ever – widening, increasingly lighter space that culminated into a great hall with a ceiling fresco. The gilt framed painting could not contain the activity of the painting as the figures over spilled the stage – like theatre painted on the ceiling. A man's legs were loosely draped over the edge of the frame while a wolfhound and soldier had fallen out on to the ledge below!

3 Mediterranean Architecture

(1) An Introduction。

Mediterranean architecture was especially popular between 1915 and 1940, and still remains one of the most popular ones there is. The Mediterranean architecture was deeply influenced by the fact that these parts include three very different continents: Europe (Spain, Greece, France, Italy and the Balkan), Asia (Turkey, Cyprus, Lebanon) and Africa (Morocco, Algeria, Tunisia, Libya), and they have had a very long and prolific history with many civilizations. It was used all over the world, for designing buildings in places like California or Florida, that were inspired by the buildings in the Mediterranean, due to the fact that they have the same climate. These structures are specific for having elements like courtyards or terraces.

第五部分　建筑文本翻译拓展训练（一）英译汉

Mediterranean houses have some elements that raise their aesthetics. For example, they have porticos supported with Doric columns, balconies with parapet walls, and colonnades with keystones on the top. There can be chandeliers and mantelpieces inside, and gardens, with decorative landscape plants, and fountains outside. The buildings can have open porches, divine towers and tall turrets, spiral staircases and a loggia (a chain of arches supported by columns). The houses can be situated next to the ocean or built in the hills. Most of the houses have window grilles made out of wood, because they have to withstand different weather conditions like rain, sun, wind and salt from the sea.

The basic thing to know here is that the type of building structure determines the way the building is built. For example, this means that an Italianate villa will be built differently than a farmhouse, which sounds reasonable enough, but is still something worth emphasizing. The design for the buildings has many origins and can be inspired by, for instance, Spanish churches or can have its origins Moorish, Tuscan or Andalusian architecture. Many homes have gates and fences to give its residents some privacy, they have more than one floor and many bedrooms, and the entrance to the building usually has a vestibule. The roof is either made from red tiles or hipped, and they have eaves under the edge. The most used colors inside are lavender, sky blue, pale yellow, and there can be swimming pool outside.

Due to the hot climate, especially during the summer when the temperatures can reach 40 degrees Celsius, houses are built so that the inside remains constantly cool. This is done by constructing thick walls. They are, therefore, built taking into account different environmental (exp. sound) and weather factors, like rain, wind, and temperatures from which the people inside should be protected from. This can be done by using insulation to keep the heat from getting inside, by giving the walls a mortar finish. And to avoid further absorption, the walls and floors are usually white so that they reflect heat.

So we could say that the buildings have both an allocation and that they look good, blending functionality with elegance. They are very open and have good ventilation, with a lot of breezy areas inside keeping in mind the warm climate of the Mediterranean implemented into the architecture.

The dominant influences of Mediterranean house plans is the architecture of Spain and Italy. "Mediterranean style house plans" is more of an umbrella term that embodies the likes of these architectural styles. Here we will briefly cover the family of Mediterranean house plans that include the Italian Renaissance, the Mission home plans, and the Spanish Eclectic style house plans.

The Italian Renaissance, also referred to as the Tuscan style, have low – pitched roofs (usually hipped) typically covered by ceramic tiles. Other features include full – length first – story windows with radius top, wide overhanging eaves supported by decorative brackets, upper story windows usually smaller and less elaborate than the windows on the main level of this Mediterranean house plan. Common decorative details include quoins, pedimented windows, classical door surrounds, molded cornices, and belt courses. Stucco or masonry walls are universal in the Mediterranean style.

The Mission style is identified by a shaped Mission dormer or roof parapet. The roof overhangs are wide and usually open or exposed and the wall surface is usually smooth stucco. Other features include porch roofs that are supported by large square piers with an arched top, and red clay tile roof covering normally seen in Mediterranean style house plans.

The Spanish Eclectic style house plans are low – pitched roofs with little or no eave overhang. Arches are prominent in windows and doors and the wall surface is normally stucco. Most Spanish Eclectic home plans are asymmetrical in its design.

(2) Mediterranean Characteristics。

The design characteristics of the Mediterranean region have evolved from its climate, topography, ancient religious buildings and people's lifestyles and cultures. The rocky, sloping sites with a view to the sea and a temperate climate have resulted in some recognizable design characteristics.

This 'style' tends to be an open plan to capture the views from as many rooms as possible. Flat roofs are common because of the infrequent precipitation, the accommodation of roof terraces and regulations to not block a neighbor's view.

For ease of construction on the sloping sites the building forms remain fairly boxy. The use of structural columns and arches are important because of the regions classical roots. A courtyard with a fountain integrates an outdoor living area used for leisure and dining with water sculpture. The water imparts the suggestion of refreshment and rejuvenation into an otherwise semi – arid climate, along with the musical quality that dripping water can impart.

Portals, loggias, porticos, verandas and terraces are all common due to the pragmatic need of shading the interior and the desire to create outdoor rooms to take advantage of the temperate climate. The transition between interior and exterior space is easy and convenient. We have come to associate these and other design characteristics as Mediterranean 'style' architecture because of their predominant use in this region.

5.2 Terms and Vocabularies

① Gothic architecture 哥特式建筑

② medieval adj. 中世纪的

③ Romanesque architecture 罗马式建筑

④ Renaissance architecture 文艺复兴时期建筑

⑤ characteristic n. 特征；特性

⑥ arch n. 弓形，拱形；拱门

⑦ ribbed vault 肋架拱顶

⑧ flying buttress 飞拱

⑨ cathedral n. 教堂

⑩ abbey n. 大修道院，大寺院

⑪ ecclesiastical adj. 教会的；牧师的

⑫ heritage n. 遗产；传统

⑬ Norman architecture 诺曼式建筑

⑭ monastery n. 修道院

⑮ gatehouse n. 门房

⑯ ambulatory n. [建] 回廊

⑰ spire n. [建]尖顶；尖塔；螺旋

⑱ tympanum n. [无脊椎] 鼓室

⑲ aesthetic adj. 美的；美学的

⑳ ideological adj. 思想的；意识形态的

㉑ pinnacle n. 高峰；小尖塔

㉒ Sassanian architecture 萨珊架构，萨珊王朝艺术是波斯艺术的顶峰

㉓ Syria n. 叙利亚共和国

㉔ Mesopotamia n. 美索不达米亚（亚洲西南部）

㉕ secular adj. 世俗的；现世的；不朽的

㉖ Muslim adj. 伊斯兰教的

㉗ Islamic architecture 伊斯兰建筑

㉘ Carolingian adj. 法国卡洛林王朝的

㉙ façade n. 外观；（法）建筑物正面

㉚ portal n. 大门，入口

㉛ congestion n. 拥挤

㉜ suffuse vt. 充满；弥漫

㉝ mason n. 泥瓦匠

㉞ chapel n. 小礼拜堂，小教堂

㉟ clerestory n. 天窗；长廊

㊱ prototype n. 原型；标准，模范

㊲ domain n. 领域

㊳ transept n. 教堂的十字型翼部

㊴ spectacular adj. 壮观的，惊人的

㊵ Angevin dynasty 安茹王朝（又常被称作金雀花王朝）

㊶ the Low Countries 低地国家（特指荷兰、比利时和卢森堡，因其海拔低而得名。）

㊷ Sicily n. 西西里岛（意大利一岛名）

㊸ revival n. 复兴；复活

㊹ barbarous adj. 野蛮的；残暴的

㊺ licentious adj. 放肆的；放纵的

㊻ monkish adj. 僧侣的；苦行僧般的

㊼ premeditate vi. 预谋；预先考虑

㊽ luxuriant adj. 繁茂的；丰富的；奢华的

㊾ scorn n/v. 轻蔑；嘲笑

㊿ epoch n. 新纪元

(51) plunge vi. 投入；陷入

(52) pavilion n. 阁；亭子

(53) Chicago Columbian Exposition 芝加哥哥伦比亚博览会

(54) plaster n. 石膏；灰泥

(55) vertical adj. 垂直的，直立的

(56) terra cotta 赤褐色的；赤土陶器

(57) ornament n. 装饰；[建][服装]装饰物

(58) stronghold n. 要塞；大本营；中心地

(59) aisle n. 通道，走道；侧廊

(60) nave n. （教堂的）中殿

(61) sanctuary n. 避难所；至圣所

(62) Triforium n. 教堂拱门上之三拱式拱廊

(63) Byzantine adj. 拜占庭式的；东罗马帝国的

(64) semi – circular adj. 半圆形的

(65) apex n. 顶点；尖端

(66) lateral adj. 侧面的，横向的

(67) transversely adv. 横着；横断地；横切地

(68) longitudinally adv. 长度上，经向；经度上

(69) pier n. 码头；桥墩；窗间壁

(70) metaphor n. 暗喻，隐喻

(71) allude vi. 暗指

(72) skeletal adj. 骨骼的

(73) pinnacle n. 高峰；小尖塔

(74) clergy n. 神职人员；牧师；僧侣

(75) span n. 跨度，跨距；范围

(76) perpendicular adj. 垂直的，正交的；直立的；陡峭的

(77) dumpy adj. 粗短的；忧郁的

(78) Notre Dame Cathedral 巴黎圣母院

⑦⑨ sexpartite adj. 分成六个的；由六部分组成的

⑧⓪ Durham Cathedral 达勒姆座堂

⑧① obsolete adj. 废弃的；老式的

⑧② Chartres n. 沙特尔（法国博斯的首府）

⑧③ Chevet n. 伸出的礼拜堂

⑧④ scaffold n. 脚手架；鹰架

⑧⑤ lancet n. 尖顶窗

⑧⑥ oculus n. 眼睛

⑧⑦ zodiac n. 黄道带，十二宫图

⑧⑧ cavernous adj. 似巨穴的

⑧⑨ Amiens Cathedral 亚眠主教座堂

⑨⓪ canopy n. 天篷；华盖；遮篷；苍穹

⑨① Colonnades n. （建筑工程）柱廊

⑨② Rococo n. 洛可可式（18 世纪后半期盛行欧洲的一种建筑装饰艺术风格）

⑨③ Baroque 巴洛克风格

⑨④ grandiloquence n. 豪言壮语，豪语；夸张之言

⑨⑤ linearity n. 线性；线性度；直线性

⑨⑥ Parisian adj. 巴黎人的；巴黎的

⑨⑦ cabinetmakers n. 家具工；细工木匠

⑨⑧ bronze n. 青铜

⑨⑨ arabesque n. 蔓藤花纹；阿拉伯式花纹

⑩⓪ scroll n. 卷轴，画卷；卷形物

⑩① ingenious adj. 有独创性的

⑩② bizarre adj. 奇异的

⑩③ chinoiserie n. 具有中国艺术风格的物品；中国艺术风格

⑩④ supersede vt. 取代，代替

⑩⑤ asymmetric adj. 不对称的；非对称的

⑩⑥ wig n. 假发

⑩⑦ flamboyance n. 华丽；炫耀

⑩⑧ frill n. 装饰；褶边

⑩⑨ Mediterranean architecture 地中海建筑

⑩⑩ turret n. 炮塔；角楼

⑩⑪ Moorish 摩尔式的建筑风格

⑩⑫ Tuscan adj. 托斯卡纳的（源自于意大利乡村式的建筑设计）

⑩⑬ Andalusian adj. 安达卢西亚的

⑩⑭ pediment n. 山形墙；三角墙

5.3　Syntactic Structure 翻译及重点句型分析

① Originating in 12th century France and lasting into the 16th century, Gothic architecture was known during the period as "the French Style" with the term Gothic first appearing during the latter part of the Renaissance.

哥特式建筑起源于12世纪的法国，并一直持续到16世纪，期间人们一直将其认为是"法式风格"建筑的代表，后来的文艺复兴时期才首次出现了哥特式建筑一词。

解析：本句的主句是"Gothic architecture was known during the period as "the French Style"，以 sth/sb be known as 的短语出现。Originating in 12th century France and lasting into the 16th century 是一个现在分词短语，做伴随状语成分。with the term Gothic first appearing during the latter part of the Renaissance 是由介词 with 引导的伴随状语部分。

② It is in the great churches and cathedrals and in a number of civil buildings that the Gothic style was expressed most powerfully, its characteristics lending themselves to appeal to the emotions.

大教堂和一些市民建筑最能展现出哥特式风格，其特色完全吸引住了人们。

解析：本句的句式是"it is...that"，是一个强调句，强调的是地点状语"in the great churches and cathedrals and in a number of civic buildings"。its characteristics lending themselves to appeal to the emotions 是一个独立主格部分，主语是 its characteristics。

③ A great number of ecclesiastical buildings remain from this period, of which even the smallest are often structures of architectural distinction while many of the larger churches are considered priceless works of art and are listed with UNESCO as World Heritage Sites.

在这段时期内保存下来的众多基督教会建筑中，即便是那些小型的建筑也能体现出其结构上的差别，而那些大型建筑则被认为是无价珍品，并被收录在联合国教科文组织列的世界文化遗产名录中。

解析：本句的主句是"A great number of ecclesiastical buildings remain from this period"，后面的"of which even the smallest are often structures of architectural distinction while many of the larger churches are considered priceless works of art and are listed with UNESCO as World Heritage Sites"是定语从句部分，关系代词 which 指代 ecclesiastical buildings。定语从句中"while many of the larger churches are considered priceless works of art and are listed with UNESCO as World Heritage Sites"一句是由 while 引导的从句。

④ Romanesque architecture, or Norman architecture as it is generally termed in England because of its association with the Norman invasion, had already established the basic architectural forms and units that were to remain in slow evolution throughout the Medieval period.

罗马式建筑，或叫诺曼底建筑，在英格兰，是一种概括性的术语，因为它和诺曼底人的入侵有关；该种建筑已经建立起了基本的建筑样式和单元，并且在整个中世纪发展缓慢。

解析:本句的主语是"Romanesque architecture, or Norman architecture""as it is generally termed in England because of its association with the Norman invasion"是由 as 引导的状语从句,主句部分是"had already established the basic architectural forms and units that were to remain in slow evolution throughout the Medieval period"。在主句中,由 that 引导定语从句,先行词为"the basic architectural forms and units"。

⑤ The widespread introduction of a single feature was to bring about the stylistic change that separates Gothic from Romanesque, and broke the tradition of massive masonry and solid walls penetrated by small openings, replacing it with a style where light appears to triumph over substance.

一个简单特色的广泛引进导致了风格上的迥异变化,使得哥特式建筑从罗马式建筑区分了出来,并打破了使用大量砖砾及带有小圆孔的固体墙的传统,取而代之的是使用那种使得光线占主要地位的建筑样式。

解析:本句的主语是"the widespread introduction of a single feature",谓语部分是由动词 was 和 broke 并列构成,其后的"replacing it with a style where light appears to triumph over substance"是动名词短语,做伴随状语。

⑥ While so-called 'pitched' brick vaulting, which could be constructed without centering, may date back in the Ancient Near East to the 2nd millennium BC, the earliest evidence of the pointed masonry arch appears in late Roman and Sassanian architecture, mostly evidenced in early church building in Syria and Mesopotamia, but occasionally also in secular structures like the Karamagara Bridge.

尽管那种所谓的"倾斜的"砖制屋顶在建造时并没有一个中心为基础,也许得回溯到古代近东时期到公元前 2000 年,但表明尖石拱顶存在的最早证据可以在后罗马和萨桑王朝建筑中发现;最明显的是在叙利亚和美索不达米亚那的早期教堂里,当然也可以偶尔在非宗教建筑如 Karamagara Bridge 中发现线索。

解析：本句中"While so-called 'pitched' brick vaulting, which could be constructed without centering, may date back in the Ancient Near East to the 2nd millennium BC"是由 while 引导的状语从句，其中"so-called 'pitched' brick vaulting"是状语从句的主语，which 引导的定语从句对主语进行修饰。"may date back in the Ancient Near East to the 2nd millennium BC"是状语从句的谓语部分。"he earliest evidence of the pointed masonry arch appears in late Roman and Sassanian architecture, mostly evidenced in early church building in Syria and Mesopotamia, but occasionally also in secular structures like the Karamagara Bridge"部分是主句部分，其中"he earliest evidence of the pointed masonry arch"是主语，"mostly evidenced in early church building in Syria and Mesopotamia, but occasionally also in secular structures like the Karamagara Bridge"部分是过去分词短语，做伴随状语。

⑦ To achieve his aims, his masons drew on the several new features which evolved or had been introduced to Romanesque architecture, the pointed arch, the ribbed vault, the ambulatory with radiating chapels, the clustered columns supporting ribs springing in different directions and the flying buttresses which enabled the insertion of large clerestory windows.

为了达到这一目标，他的工匠用了一些曾被引进到罗马式建筑中的新特征，如尖拱顶，扇形拱顶，辐射状的回廊，集柱以支撑不同方向的弯梁起拱石和飞扶垛，正是这些使得那些大型天窗能够嵌入进去。

解析：本句中"to achieve his aims"是不定式做目的状语，主语是 his masons，谓语动词是 drew on，宾语由一系列名词短语构成，"the several new features, he pointed arch, the ribbed vault, the ambulatory with radiating chapels, the clustered columns and the flying buttresses"，在名词短语中部分短语有定语从句修饰，如 "the several new features which evolved or had been introduced to Romanesque architecture" 和 "the flying buttresses which enabled the insertion of large clerestory windows" 都是由 which 引导的定语从句。

⑧ For the first time, an attempt was made to destroy an instinctive and, so far as Europe was concerned, an almost universal form of art, and to substitute in its place another built up by artificial rules and premeditated theories; it was necessary, therefore, that the ground should be cleared of a once luxuriant growth that still showed signs of vitality, and to effect this the schools of Vignola, Palladio, and Wren were compelled to throw scorn on the art they were determined to discredit.

对于欧洲人来说，这是史上第一次企图破坏一种本能的，几乎是普遍被大家所接受的艺术形式，取而代之的是由虚伪的原则和预先设定好的理论所建立起的艺术形式。因此，这种艺术的根基的确立有必要先清除那种曾经繁荣一时，且生命力尚存的艺术形式。为此，Vignola，Palladio 和 Wren 学派不得不对这一他们注定要怀疑的艺术表示轻视。

解析：本句由连个分句组成，并由分号连接。第一个分句的主语是 "an attempt"，谓语部分由 "was made to destroy…" 和 "and to substitute…" 共同构成。第二个分句由 "it was necessary, therefore, that the ground should be cleared of a once luxuriant growth that still showed signs of vitality" 和 "and to effect this the schools of Vignola, Palladio, and Wren were compelled to throw scorn on the art they were determined to discredit" 两个并列句组成。

⑨ As ignorant of the true habitat of the style as they were of its nature, the Italians of the Renaissance called it the "maniera Tedesca", and since to them the word Goth implied the perfection of barbarism, it is but natural that they should have applied it to a style they desired to destroy.

由于对这一艺术形式来源的无知，文艺复兴时期的意大利人称之为"maniera Tedesca"，同时因为"哥特"一词对他们来说意味着野蛮，他们很自然地将这一艺术形式归为应该被消灭的一类。

解析：本句由"the Italians of the Renaissance called it the 'maniera Tedesca'和"and since to them the word Goth implied the perfection of barbarism, it is but natural that they should have applied it to a style they desired to destroy"两个并列句组成。第一个句子中"As ignorant of the true habitat of the style as they were of its nature"是状语部分。第二个句子的主句是"it is but natural that they should have applied it to a style they desired to destroy" " since to them the word Goth implied the perfection of barbarism"是由 since 引导的原因状语从句。

⑩ The style ceased, for the particular type of civilization it expressed had come to an end; but the name remained, and when, early in the nineteenth century, the beginnings of a new epoch brought new apologists, the old title was taken over as the only one available, and since then constant efforts have been made to define it more exactly, to give it a new significance, or to substitute in its place a term more expressive of the idea to be conveyed.

由于它所代表的文明走到了尽头，这一艺术形式也结束了，但名字仍然保留着，直到十九世纪早期，历史新纪元为这一名词带来了新的辩护者，旧名字被取代，而后人们做出大量努力使这一名称更准确，赋予它更重要的意义，并试图用一个更能表达其想法的名称替代旧名称。

解析：本句由两个分句组成，并由分号连接。第一个分句是的主句是"The style ceased"，"for the particular type of civilization it expressed had come to an end"是由 for 引导的原因状语从句。第二个分句由三个并列句构成。

⑪ Whereas Norman architecture can be seen as being 'dumpy' due to their more limited knowledge of building, the Gothic era coincided with a greater knowledge of engineering and this is reflected in the church buildings completed during this era.

虽然诺曼式建筑风格因其建筑知识有限而被认为"破败",哥特时期却是工程知识大发展的时期,这一点在这一时期的教堂建筑中得以体现。

解析:本句是主从复合句,"Whereas Norman architecture can be seen as being'dumpy'due to their more limited knowledge of building"是由 whereas 引导的状语从句,主句是"the Gothic era coincided with a greater knowledge of engineering and this is reflected in the church buildings completed during this era"。

⑫ That is why it worked best with small scaled items, which differed greatly to the large Baroque sculpture and architecture of previous eras.

这也是为什么这种艺术风格通常出现在小范围的设计中,与大型的巴洛克雕塑和以前的建筑风格截然不同。

解析:本句是表语从句,由 why 引导。Which 引导的宾语从句的先行词是 small scaled items。

⑬ Around 1670, King Louis XIV (14th) decided to turn a royal hunting lodge, a few miles outside of Paris in Versailles, into a great palace.

1670 年左右,国王路易十四决定将一个位于离巴黎几英里以外的凡尔赛的皇家狩猎小屋改造成一个恢宏的宫殿。

解析:本句是一个结构复杂的单句,主语是"King Louis XIV (14th)",谓语部分是"decided to turn a royal hunting lodge into a great palace",其中"a few miles outside of Paris in Versailles"是宾语"a royal hunting lodge"的补足语。

⑭ It was used all over the world, for designing buildings in places like California or Florida, that were inspired by the buildings in the Mediterranean, due to the fact that they have the same climate.

地中海风格被全球所接受,比如加利福尼亚和弗罗里达的建筑就受到地中海地区建筑的影响,因为它们的气候很相似。这些建筑结构的具体特点是,有庭院和阳台等元素。

解析:本句是一个结构复杂的单句,主语是"it",谓语部分是"was used all over the world"。"for designing buildings in places like California or Florida"和"due to the fact that they have the same climate"是状语部分,引导原因和目的。

⑮ They are, therefore, built taking into account different environmental (exp. sound) and weather factors, like rain, wind, and temperatures from which the people inside should be protected from.

所以,房屋在建设时已经考虑到了不同的环境和气候因素,像雨、风和室内温度等。

解析:本句是一个结构复杂的单句,主语是"they"。"taking into account different environmental (exp. sound) and weather factors, like rain, wind, and temperatures from which the people inside should be protected from"是伴随状语部分,其中which引导的定语从句修饰先行词"weather factors, like rain, wind, and temperatures"。

5.4 参考译文

哥特式、洛可可式及地中海式建筑

1. 哥特式建筑

(1) 哥特式建筑简介。

哥特式建筑指的是一种建筑风格,在欧洲中世纪后期比较兴盛。它是从罗马式建筑演变而来的,后来的文艺复兴式建筑继承了其风格。

哥特式建筑起源于 12 世纪的法国，并一直持续到 16 世纪，期间人们一直将其认为是"法式风格"建筑的代表，后来的文艺复兴时期才首次出现了哥特式建筑一词。其主要特征包括尖顶、圆形拱顶和飞拱。

我们所熟悉的哥特式代表建筑有欧洲的大教堂，修道院和教堂。很多城堡、宫殿、市政厅、公馆、大学以及相当一部分的私人会所也采用了哥特式风格。大教堂和一些市民建筑最能展现出哥特式风格，其特色完全吸引住了人们。在这段时期内保存下来的众多基督教会建筑中，即便是那些小型的建筑也能体现出其结构上的差别，而那些大型建筑则被认为是无价珍品，并被收录在联合国教科文组织列的世界文化遗产名录中。因此可以说，对哥特式建筑的研究主要是集中在对大教堂的研究上。

（2）哥特式建筑的历史。

哥特式建筑是从先前的罗马式建筑流派发展而来的。大多数情况下，它们并没有明显的分界，后来 Brunelleschi 提倡古典建筑的复兴，演变为 15 世纪佛罗伦萨文艺复兴式建筑。

罗马传统

罗马式建筑，或叫诺曼底建筑，在英格兰，是一种概括性的术语，因为它和诺曼底人的入侵有关；该种建筑已经建立起了基本的建筑样式和单元，并且在整个中世纪发展缓慢。大教堂、教区教堂、修道院、城堡、宫殿、大礼堂、小门房等的基本结构都已建立起来。基督教会建筑的一些主要特色包括扇形拱顶、扶墙、集柱、回廊、轮形扇窗、尖顶及雕刻丰富的墙壁三角面。

一个简单特色的广泛引进导致了风格上的迥异变化，使得哥特式建筑从罗马式建筑区分了出来，并打破了使用大量砖砾及带有小圆孔的固体墙的传统，取而代之的是使用那种使得光线占主要地位的建筑样式。带来这种改变的特色则是尖拱顶的使用。它的使用迎来了其他建筑手段的发展，先是将其在稀疏型建筑中作试验，接着将其应用到一些公共机构中，以满足新样式结构上，艺术上和意识形态上的需求。这些哥特式基督教会建筑典型特征包括飞拱，尖顶及有图样的窗户。

可能受到的东方影响

尽管那种所谓的"倾斜的"砖制屋顶在建造时并没有一个中心为基础，也许其得回溯到古代近东时期到公元前2000年，但表明尖石拱顶存在的最早证据可以在后罗马和萨桑王朝建筑中发现；最明显的是在叙利亚和美索不达米亚那的早期教堂里，当然也可以偶尔在非宗教建筑如Karamagara Bridge中发现线索。在穆斯林统治的第7世纪之后，它逐渐成为伊斯兰建筑的一个标准特征。

有理论认为说，1090年罗曼第人对Islamic Sicily的征服，1096年十字军东征以及西班牙穆斯林的存在增进了与穆斯林世界的军事和文化交流，为中世纪的欧洲带去了尖拱顶这样的新知识。另有理论认为尖拱顶在西欧是自然发展而来的，它是一个纯粹的建筑结构技术性问题解决方案，同时也是法国和英国教堂早期引进使用的一种风格特征。

Abbot Suger

Abbot Suger是法国国王路易斯六世和路易斯七世的知心朋友，他于1137年决定重建Saint-Denis大教堂，该教堂和一个同为皇家住宅的修道院相连。

Suger从西面开始，将原来的加洛林式的正面单扇门进行了改建。他将Saint-Denis教堂的正面设计成一种对加洛林罗马拱顶的对应样式，将其划分为三个部分，有三块大门，以缓解拥挤。法国西门的玫瑰窗是已知最早的样本。

1140年，西门建成，Abbot Suger着手东门建设，加洛林的中殿可以用了。他设计了一个可以让光线弥漫的高坛。为了达到这一目标，他的工匠用了一些曾被引入罗马式建筑中的新特征，如尖拱顶，扇形拱顶，辐射状的回廊，集柱以支撑不同方向的弯梁起拱石和飞扶垛,正是这些使得那些大型天窗能够嵌入进去。

该新建筑在1144年6月11号完成,国王出席并题词。Saint-Denis修道院因此成为日后北法皇家地区建筑的原型。通常其被认为是第一座哥特式建筑。100年之后,原先Saint-Denis教堂的中殿以哥特式风格被重建,在中殿左右翼增加了两个玫瑰窗。在金雀花王朝时期,这种风格被引入英国,并传播到了法国,低地国家(荷兰、卢森堡、比利时),德国,西班牙,北意大利和西西里岛。

(3) 哥特式建筑的复兴。

在18世纪中期的英国,哥特式建筑开始复兴,19世纪扩大到整个欧洲,并一直持续到20世纪,主要体现在教会和大学里的建筑物上。

然而,哥特一词最开始时被用作轻蔑的意思。Vasari说过,"然后出现了一群新建筑师,他们遵循自己野蛮国家的习惯,建起一些被称为哥特式的建筑",而Evelyn则在写作中表达了他这一时期的观点,"古希腊和罗马建筑为所有完美无缺的伟大建筑制定了标准"——但哥特人和汪达尔人破坏了这些规矩,同时"建立起一种空想的,放纵的建筑风格:一堆沉重,灰暗,令人忧伤,苦行僧般的建筑,毫无比例和实用性,也无美感可言。"对于欧洲人来说,这是史上第一次企图破坏一种本能的,几乎是普遍被大家所接受的艺术形式,取而代之的是由虚伪的原则和预先设定好的理论所建立起的艺术形式。因此,这种艺术的根基的确立有必要先清除那种曾经繁荣一时,且生命力尚存的艺术形式。为此,Vignola,Palladio和Wren学派不得不对这一他们注定要怀疑的艺术表示轻视。由于对这一艺术形式来源的无知,文艺复兴时期的意大利人称之为"maniera Tedesca",同时因为哥特一词对他们来说意味着野蛮,他们很自然地将这一艺术形式归为应该被消灭的一类。由于它所代表的文明走到了尽头,这一艺术形式也结束了,但名字仍然保留着,直到十九世纪早期,历史新纪元为这一名词带来了新的辩护者,旧名字被取代,而后人们做出大量努力使这一名称更准确,赋予它更重要的意义,并试图用一个更能表达其想法的名称替代旧名称。

值得注意的是,哥特式建筑的复兴并没有排除其他所有建筑风格。哥特式建筑只是 19 世纪风行的几种复兴建筑形式之一。在 G. A. Audsley 移民的 19 世纪 90 年代早期,他应该见到过许多罗马建筑的复兴。这一风格由 Henry H. Richardson 最先在美国倡导。Henry H. Richardson 深受法国南部历史建筑的影响。"理查德森式"罗马建筑充满大量大块切割的石头,对比强烈的色彩以及纪念碑式的圆拱。这一建筑形式随着 1893 年的芝加哥哥伦比亚博览会上象征罗马建筑复兴建筑的建立而成为国家级建筑形式。考虑到 G. A. Audsley 对 19 世纪 90 年代罗马建筑复兴的恶评,他可能对这一建筑形式不感兴趣。

事实证明,哥特式建筑复兴比罗马建筑复兴更长久。19 世纪末出现了一些主要教堂建筑,包括纽约、华盛顿和利物浦,在这些地方中世纪哥特式建筑结构系统得以复兴。在这些地方的建筑中,不允许使用早期建筑中的石膏拱顶和铁柱子,大部分建筑全部是石头构成,而没有钢架。

令人惊奇的是,哥特式建筑在包含大量钢结构的 20 世纪建筑中也十分流行。美国建筑师发现了哥特式建筑的强调垂直的特点,并运用于摩天大厦中,其中最有名的是 Cass Gilbert(1859~1934)在纽约建的伍尔沃斯大楼(1943)。这座 880 英尺高的建筑物外面由赤褐色的嵌板所构成,上面嵌有哥特式装饰物,并由钢架支撑。

到 1930 年为止,哥特式建筑复兴一直按常规发展。作为中世纪哥特式建筑的最后堡垒,教堂和大学校园庇护了哥特式建筑复兴的尾声。同为未来派风格的装饰艺术和现代主义最终取代了传统艺术。

(4) 哥特式建筑形式。

平面:典型的平面是 3 或 5 侧廊的十字型平面,这就意味着中殿在每边都有 1~2 个走廊。平面包括十字型翼部、圣乐坛和至圣所。

在哥特时期，回廊是用来张贴通告的处所，并附有小教堂。西侧是主入口，通常有一或两个纪念碑式的塔楼。

室内立面图：像许多罗马教堂一样，哥特式教堂内部被拱廊、三拱式拱廊和长廊所分割。早期教堂在拱廊和三拱式拱廊间也有走廊，但后来由于长廊的窗户越来越大而被取消。

尖顶：与早期基督教堂，东罗马教堂和罗马教堂的半圆拱不同，中世纪晚期的建筑师设计哥特式尖顶有许多原因。罗马式半圆拱可以完美地俯瞰下面的广场，但由于拱顶的顶端高度不同，不能覆盖矩形空间。而尖顶可以修的更扁或更尖而不那么明显，所以，尖顶可以更优雅地覆盖矩形或方形空间。现在的建筑大师可以很容易地设计矩形或方形预配图，同时，尖顶比圆顶所产生的侧向推力更小，所以更容易支撑。总的来说，尖顶比巨大的、坚实的圆顶看起来更轻便，更精致。

肋架拱顶：当尖顶结合在一起形成拱顶，拱顶由于结构强度增大而呈肋状。肋架是由尖拱呈对角线（穿过中殿），横切（与中殿垂直）或纵切（与中殿平行）组合而成。这些结构肋架使得肋架间的填充板可以由重量较轻的石头充当，以减轻整个拱顶的重量。这些扇形肋穹顶和飞拱共同使拱顶的作用力集中在外部支柱上，使得墙上可以开窗户，进阳光。哥特式教堂的框架很像骨架。扇形肋穹顶即是一个有机（自然地）比喻（比较），暗指人身体上的肋骨。

飞拱：通过连接被称为飞拱的桥梁，中殿拱顶的侧向力被转移到了建筑物外部的支柱上。飞拱飞过走廊到达外部支柱。尖塔安置在飞拱的顶端以减轻重量和额外的压力。这种新的哥特式建筑发明使得飞拱间的墙上可以布满窗户。

石雕花格窗和彩色玻璃：随着结构知识的进步，哥特式建筑的设计师们试图在建筑物内创造精神力量。由石雕花格窗连接的满是彩色玻璃的半透明墙就像一面布满蕾丝的墙，墙上描绘着圣经里的图形和场景。

雕刻：教堂外部由石质雕刻覆盖，描述了圣经里提到的国王和其他贵族，以及重要的教士和故事。精致的石质雕刻表现了石头和装饰物间的精巧平衡。这种艺术的结合是如此完整，以致于将设计、结构和艺术都融为一体。

（5）哥特式教堂建筑。

中世纪英国的哥特式教堂建筑发展于诺曼式建筑。"哥特式建筑"一词指的是1200～1500年的建筑风格。如此大的时间跨度意味着在哥特式建筑中还发展出一些其他建筑风格，我们通常将这些建筑风格划分为三类。1200～1300年的建筑通常指的是英国早期建筑；1300～1400年的建筑指盛装建筑；1400～1500年的建筑则指垂直式建筑。我们很容易看到一些大教堂呈现出这些时期的特征。

哥特式建筑的突出特点是大型塔楼和尖顶。虽然诺曼式建筑风格因其建筑知识有限而被认为"破败"，哥特时期却是工程知识大发展的时期，这一点在这一时期的教堂建筑中得以体现。

哥特式教堂从根本上与诺曼式建筑不同。多年来学到的知识和技能意味着石头可以被雕刻成特殊形状，从而可以精确地与周围的石头接缝。因此，诺曼式建筑所喜欢的大石头被有特殊形状的石头所替代。另一个主要变化在于诺曼式建筑所采用的空墙不再被后来的建筑师所使用。墙和柱子都是实心的，使得它们可以承受更大的重量。这些简单的事实使得教堂，尤其是大教堂比诺曼时期的教堂大得多。这一特点同教堂筹钱的能力一并解释了为什么哥特时期教堂比以前的教堂要大。

尖顶的采用也促进了教堂建筑的发展。正是这种形状比诺曼时期的圆顶能承受更多重量。现在的教堂屋顶比诺曼时期大得多，因此，他们更重。为了确保墙和支柱能支撑这些重量，这一时期的建筑师发明了我们所知的飞拱。这些飞拱是教堂主体的附加部分，功能在于将额外重量转移到教堂的附加部分，而不是直穿中殿到达基座。建筑师将重量分散到建筑物中的其他支点上。飞拱抵挡了巨大屋顶的外来压力。

人们担心约克大教堂屋顶的重量是因为除了走廊的拱顶，其他拱顶都是木质的，虽然能够减少对支柱，基座等的压力，但也导致了未来火灾和蚤死虫的问题。约克大教堂的确有飞拱，但都是在十九世纪才加上去的。

能够处理更重的结构使得哥特式建筑师们可以使用更大的窗户。诺曼时期人们只能采用狭缝窗户，而现在的窗户则可以采用大型彩色玻璃窗。约克大教堂的大东窗有一个网球场大，这在诺曼时期是不可想象的。

这些新的大建筑非常昂贵。那么教堂从哪儿弄到这么多钱？钱基本上来源于英国人。农民和城市居民向教堂付很高的税—洗礼税，结婚税和死亡税；十一税，以及几百年来人们不得不免费为教堂耕作。这些税收负担了大教堂的建设，如林肯、约克、坎特伯雷和奇切斯特等处的大教堂。

（6）著名的哥特式大教堂。

巴黎圣母院，巴黎，1152~1420 年

巴黎圣母院是保守和进步思想相结合的早期哥特式建筑。平面最开始是五条带有四个拱顶的走廊跨过侧面走廊，及六个拱顶跨过中殿所有方形走廊。整体布局从一边一直延续到另一边，侧面走廊尽头是双向回廊，期间只有几处交叉通道，几乎不会破坏外立面线条。中殿、圣坛和交叉甬道是完全不同的空间，但流畅的过渡使它们能够和谐地平衡。与达勒姆大教堂不同是，巴黎圣母院没有支柱和复合支柱的替换系统，而只有大柱子支撑一些连接在一起的小柱子，这些小柱子一直延伸到拱顶的顶部。

1250 年，巴黎圣母院被"现代化"了，来跟法国其他新生代大教堂设计竞争。在甬道的侧立面，巨大的玻璃上嵌有石雕花格窗和彩色玻璃。也是在这个时代，巨大尖顶被设计在了交叉处。中殿的墙也被翻新了。

这四重提升（拱廊、走廊、三拱式拱廊和长廊）已经过时。这一时期，走廊（侧廊上的一段路）也被去除了。为了使教堂采光更好，飞拱被扩大，这样长廊的窗户就可以放大。建筑物正面使垂直结构与水平因素相平衡，制造出稳定的感觉。建筑物西面被称为皇家走廊，是第一个建在教堂正面的走廊。雕塑位于北面和南面耳廊的正面，代表的是法国国王和皇室，描述了圣经上的故事。巴黎圣母院是第一个真正宏伟壮观的大教堂，成为了后来法国教堂的典范。

沙特尔大教堂，沙特尔，1190～1220 年

沙特尔大教堂完工于1220年，被认为是第一个高大的哥特式教堂，也是第一个从设计初就计划修建飞拱的教堂。正如人们从空中看到的一样，伸出的礼拜堂旁边的飞拱（在教堂的西侧）用它稳定的手臂支撑了教堂内的拱顶，就像一副脚手架。

教堂的设计是全新的组织结构。早期哥特式教堂正方形的侧厅被长方形的中殿和耳厅所替代（侧廊仍有正方形侧厅），这也成为了高大的哥特式教堂的典型特征。这一类型教堂的拱顶由四部分组成，只有四个镶板并列构成了一系列连续的空间。中殿是一个宽敞的连续厅，周围环绕的不是沉重的柱子，而是许多面彩色玻璃墙。

拱廊和长廊不需要分开的走廊就被扩大了。长廊上有2个大厅窗户（一对尖顶窗），顶上有窗孔（花型窗户）。长廊内射进来的日光由于穿过了彩色玻璃而变得暗淡而有色彩。

教堂西门有1150年的雕刻，讲述着基督的故事：基督诞生（左边），基督升天（右边），基督复活（中间）。一根根柱子上也刻有连续的基督生活的场景。其他的雕刻代表了人文学科，十二宫的标志，以及各月和各季节的劳动景象。国王和王后也刻在了石柱上。尖顶是不同时期修建的，所以有些许不同，解释了他们间的不对称。三个入口都很大，是洞穴式的三开门。

即使在教堂关门后，参观者也能从教堂外面的雕刻和浮雕上感受到宗教对人的激励和教育。教堂开放时，嵌板上的彩色玻璃也讲述着圣经的故事。总体上，沙特尔大教堂不仅是人们寻求宗教精神的地方，也是人们领受宗教教育的所在。

亚眠主教座堂，亚眠，1219~1269 年

亚眠主教座堂是法国最高的大教堂，其内部空间是高大教堂中最宽敞的。亚眠主教教堂改进了沙特尔大教堂的结构。和以往的教堂相比，它的中殿更高更细，拱廊也更高。更细的支柱比例给人以宏伟壮观之感，从拱顶到地面，形成了连续的视觉联系。圣坛的四部分拱顶好似悬垂在巨大柱子上的华盖和帐篷。光线从长廊射进来，产生了垂直的空间感，也模糊了结构的轮廓。结构技术和光削弱了物理物质。

教堂的设计是长方形开间，4 个拱肋和一个撑墙，使得更充足的光线可以照进来，而且更精确美观。这是一个真正的骨架结构。柱子间的墙只用来支撑窗户，同时抵抗坏天气对教堂的侵袭。石顶的上面是高达 61 米的木质有斜坡的桁架屋顶，既方便于除雨雪，又容易鼓满风。这样一来，就需要两套飞拱将屋顶和中殿拱顶的力转移到外部扶壁墩上。低处的飞拱转移了中殿拱顶的向外推力，而高处飞拱则将屋顶的风负载转移到撑墙上。

由于矮塔建于 14 世纪，而高塔建于 15 世纪，所以教堂正面的塔是高低不等的。入口是凹陷的，在教堂前平面的后面。由于教堂外部满是柱廊，拱门，小尖塔，玫瑰窗和其他石雕工艺，所以高墙和塔的表面几乎全部是装饰图案，而这些装饰也几乎使得人们忽略了建筑物本身。教堂主入口与长廊和中廊并齐。

2. 洛可可建筑

（1）洛可可建筑简介。

作为一种建筑形式，尤其在是室内建筑和装饰艺术方面，洛可可建筑发源于法国，于 18 世纪广泛盛行于欧洲。洛可可一词有可能源于法语 rocaille 和 coquille（石头和贝壳），指的是意大利巴洛克室内建筑和庭院建筑中较为突出的自然装饰物。洛可可的表达方式最初出现在摄政时期建筑风格中。与巴洛克夸张风格不同，洛可可是一种精巧的线性艺术。Juste Aurèle Meissonier 和 Nicholas Pineau 通过他们的雕刻使这一艺术形式传遍欧洲。巴黎的挂毯织工，家具工人和铜器工匠都追逐着这一潮流，并将蔓藤花纹元素、贝壳、画轴、树枝、花和竹子的茎等基本图案改造成独具匠心，迷人的组合图案。对中国艺术的热衷也为这一艺术形式的异域风格增加了中国图案。在法国，洛可可风格的主要倡导者是 Watteau，Boucher 和 Fragonard 及建筑师 Robert de Cotte，Gilles Marie Oppenord，以及后来的 Jacques Ange Gabriel。洛可可风潮传入德国和奥地利后，François de Cuvilliès 是当时的先锋派人物。意大利洛可可风格，尤其是威尼斯洛可可风格，体现了完美的装饰风格，在 Tiepolo 的画中得以彰显。Thomas Chippendale 的家具则证明了他在英国的影响力。在17世纪60年代~70年代间，洛可可风格与另一种更加古典的建筑风格形成了竞争关系，而后者随着路易十六登基而获胜。

洛可可建筑风格继巴洛克建筑之后出现在法国18世纪20年代，主要在室内建筑，绘画和装饰艺术上。繁复的巴洛克设计让步于轻巧的设计元素，如更多的曲线和自然图案。洛可可风格所体现的精巧和俏皮经常被认为是路易十六政权奢侈生活的反映。18世纪30年代，洛可可建筑在法国达到了顶峰。这种风格从建筑和家具设计传播到绘画和雕塑上。洛可可风格保留了巴洛克艺术中的繁复造型和错综图案，在这方面，洛可可艺术风格开始融合包括东方设计和不对称组合在内的各种不同特点。

在1700~1750年的欧洲和受欧洲影响的国家里，无论男女最流行的时尚是宽大的裙装剪影，追随着17世纪80~90年代的高大，瘦长的外形。有身份的男士要戴假发，通常是白色的；天然的头发上喷上香粉，以看起来时髦。

这一时期的区别在于宫廷里全副武装的正式装束和平时的随意打扮。随着时间的推移，需要正式装束的场合越来越少，最后在世纪末彻底消失。

洛可可艺术是这一时代的反映。它跟巴洛克风格产生，以女性化的曲线，繁复的设计和绚丽的色彩著称。相对于沉重的暗色巴洛克风格，洛可可建筑风格更轻快，更具错综的图形，并采用亮色。这种风格是对这个轻佻、快乐、闲来无事的时代的反映。

洛可可风格是一种法国建筑风格，也的确在别的国家没有像在法国那么备受推崇。这种建筑风格的灵感来自于风、花、贝壳、大海、珊瑚、画轴、喷雾等自然元素。这一时代建筑中所采用的色彩大部分是柔和亮丽的颜色。

在现存的这一时期的代表性建筑中，有马提农官邸，爱丽舍官和由 Jacques Ange Gabriel 设计的路易十五广场，也被称作协和广场。

除了错综复杂的设计和琐碎的细节，洛可可建筑还对建筑艺术作出了改进；下水道设施得到了改善，烟囱效率更高，房间的组织更合理，私密性更好。

（2）洛可可艺术的影响。

洛可可运动影响了其他的艺术形式，包括绘画、建筑和雕塑。它俏皮的本质，精巧的笔触和女性化的风格还极大的影响了家居设计、挂毯、室内设计和服饰。

洛可可主题轻快，在装饰艺术中，洛可可式作品都非常繁复。这也是为什么这种艺术风格通常出现在小范围的设计中，与大型的巴洛克雕塑和以前的建筑风格截然不同。洛可可风格最适合运用在室内设计中，如瓷器造型、花边、金属器具和家具等。

在玛丽安东瓦涅特时代，洛可可式的建筑、室内设计，甚至是服饰都十分常见。在她的宫殿里，我们可以看到房间整个都是洛可可风格的，其本身就是艺术品。洛可可式的家具装饰性很强。挂毯、镜子、装饰物和统一风格的绘画使得洛可可建筑风格更加完整。在洛可可艺术晚期，这种风格大多遭到贬损，被认为太过轻佻。

（3）巴洛克艺术与洛可可艺术。

法语"baroque"的意思是奇异或幻想。空间被像塑料和泥土一样巧妙地处理，建筑元素被有意地曲解，自然规律经常被忽视。巴洛克建筑师完成了彻底的艺术解放。然而，虽然巴洛克建筑表达了非同寻常的创造力和原创性，它也代表了极度的腐朽和过渡的装饰。洛可可风格是巴洛克风格的最后阶段。在18世纪早期，法国和德国的皇室和教会要求更庄重与奢侈的建筑风格。洛可可艺术的特点是色彩亮丽，而巴洛克艺术色彩更暗淡，感觉更沉重。洛可可建筑外观比巴洛克建筑更精巧、细腻。总的来说，洛可可时期是一个极度装饰，炫耀财富和权利的时期。

（4）洛可可建筑代表。

凡尔赛宫，法国，1670~1720年

1670年左右，国王路易十四决定将一个位于离巴黎几英里以外的凡尔赛的皇家狩猎小屋改造成一个恢宏的宫殿。一大群建筑师、雕塑家、画家和景观建筑师被召集在一起建造这个法国巴洛克/洛可可风格的伟大建筑。凡尔赛宫是人类力量掌控大自然脆弱和无序的象征。

要不是因为处于大花园一般的景观背景下，这个巨大的宫殿看起来令人无法忍受。由 Andre Le Notre 设计的花园将整个森林改造成一个集轴向排列花园，景观和小路于一体的整体。花园制造出一种从正式建筑到野生自然的逐渐过渡。宫殿旁的花园紧邻几何单元而远离宫殿本身，树林成为屏障，远处的乡村构成画卷。没有相片可以描述那种感觉，只有亲身走过花园才能体会。花园虽然是人设计的，但景观艺术却是大自然的杰作。花园随着季节和时间以及光纤的变化而变化。

主教官邸，乌兹堡，1720~1744年

洛可可时期德国最有名的建筑师是 Johann Balthazar Neumann。巴洛克和洛可可艺术在德国南部最为盛行。舞台幻术、绘画、雕塑和建筑的结合，共同体现了人们对宗教的热望和极度的财富。

Neumann 是一位将光线通过不可见光源带入建筑物的大师，使得光线感觉像是天堂之光。在位于乌兹堡的王子主教官邸，主楼梯，接待大厅和小礼拜堂内部是人们关注的重点。Neumann 提出了将楼梯从卑微的功能性元素提升到观赏性元素的观点。楼梯开始是一条单一的中心楼梯，然后延伸到反方向变成双翼，包围着中心楼梯。拾阶而上，楼梯越来越宽阔，视野越来越明亮，最后通达一个带屋顶壁画的大厅。镀金框的画无法涵盖绘画本身的动态，因为屋顶的画上人物伸出了舞台一样的戏院，一个男人的双腿搭在画框的边缘上，一只猎狼犬和士兵甚至掉到了画下面！

3. 地中海式建筑

（1）简介。

地中海式建筑在 1915～1940 年尤其风行，到如今仍然是最流行的建筑风格之一。影响地中海式建筑风格的因素之一是，地中海地区由三个大路组成：欧洲（西班牙，希腊，法国，意大利和巴尔干地区），亚洲（土耳其，塞浦路斯，黎巴嫩）和非洲（摩洛哥，阿尔及利亚，突尼斯，利比亚），这些地区有着悠久而丰富的历史和文化。地中海风格被全球所接受，比如加利福尼亚和弗罗里达的建筑就受到地中海地区建筑的影响，因为它们的气候很相似。这些建筑结构的具体特点是，有庭院和阳台等元素。

地中海式建筑有一些元素提升了它的美感。比如，有多立克柱支撑的门廊，带护栏的阳台和顶端有楔石的廊柱。室内有枝形吊灯和壁炉台，室外的花园里有景观植物装饰和喷泉。建筑物有户外门廊，神圣的塔楼和高塔以及旋转楼梯和凉廊（一排由柱子支撑的拱桥）。房子可以建在海边或山上，大多数的房子有木质窗栅，以抵御不同天气，如阴雨天、晴天、大风天和海水中的盐度。

我们应该了解的最基础的一点是建筑结构决定了建筑方式。比如，这就意味着一所意大利别墅应该建的不同于一所农舍，这虽然听着是这个道理，但仍然值得一提。建筑设计有不同的来源，有的灵感来自于西班牙教堂，有的发源于摩尔、托斯卡纳或安达卢西亚建筑。许多房屋有大门和栅栏以保护隐私，室内有好几架楼梯和多个卧室，还有门廊通往大门。屋顶铺着红瓦，或是斜脊，并在边缘有屋檐。室内最常用的颜色是薰衣草色、天蓝色和淡黄色，室外还可能有游泳池。

由于天气热，尤其是在夏天有可能达到摄氏 40 度，房屋建设时要使得室内保持凉爽，这就要求建厚墙。所以，房屋在建设时已经考虑到了不同的环境和气候因素，像雨、风和室内温度等。我们可以通过加装隔离层和砂浆避免热气进入室内，同时为了进一步防止吸热，墙和地面经常是白色的，以反射热量。

所以我们可以说地中海式建筑是集优雅和功能于一体的建筑风格。建筑开放，通风良好，同时考虑到地中海地区的温度，房屋设有多处通风点。最有影响力的地中海式建筑在西班牙和意大利。"地中海风格房屋计划"更像是一个包含了对这种建筑的喜好的涵盖性术语。在此我们简单提及一下这项计划，主要包括意大利文艺复兴、布道家庭计划和西班牙折衷主义风格计划。

意大利文艺复兴建筑，也被称作塔斯干式建筑，有缓斜的屋顶（通常是斜的），被瓷砖覆盖。其他特点还包括标准长度的有圆顶半径的落地窗，有装饰架支撑的宽大的挑檐，其上层窗户比地中海房屋计划的窗户小，而且更简洁。普遍的装饰包括房屋外墙角，人字形窗户，经典的门环绕物，飞檐成型及带状甬道。灰泥或石质的墙在地中海建筑风格中很普遍。

布道风格可以从屋顶采光窗和屋顶栏杆中识别。挑檐很宽大，或开或关，墙体表面通常是平滑的灰泥。其他特点有方形柱子支撑的玄关，其顶部是拱形以及常见的红粘土屋顶。

西班牙折衷主义风格计划有缓斜的屋顶，很少有或没有挑檐。窗户的拱顶很明显，门和墙体表面一般是灰泥。这类建筑大多不对称。

(2) 地中海建筑特点。

地中海地区的设计特点主要来自于它的气候、地质、古代宗教建筑和人们的生活方式及文化。地质多石，地势倾斜，举首可见大海，这些因素都导致了可识别的建筑特色。

这种"风格"更倾向于一种开放计划，可以使更多的房间能看到风景。由于雨水不多，同时为了不挡邻居的视野，所以平屋顶很常见。

为了方便于在斜坡上建房子，建筑物外形多为四方形。由于地区的传统，结构性支柱和拱门在建筑中很重要。带喷泉的庭院通过水的雕塑将外部生活区整合为休闲和就餐的场所。有音乐质感的滴水赋予了本来半干燥的气候以青春焕发和神清气爽的意味。

为了给室内遮阳，同时制造户外乘凉的空间等实际考虑，大门、凉廊、门廊、游廊和露台都很常见。室内外空间的转换很容易也很便利。由于这些建筑元素在这一地区被主要应用，我们通常会将它们与地中海式建筑联系起来。

简明建筑英语
翻译教程
Chapter 6

第六部分　建筑文本翻译拓展训练（二）
汉译英
——浪漫主义及功能主义建筑

6.1 文本原文

浪漫主义及功能主义建筑

1. 浪漫主义

浪漫主义推崇未被改造的自然与如画的风景,强调历史和异国风情,是对秩序和规则的反叛。浪漫主义者偏爱不对称和不规则。因其粗陋、荒凉、支离破碎、废墟整体而言成为浪漫主义痴迷的对象,展现了人类在面对大自然的无力感和自己的劳作经受岁月磨蚀的忧郁情绪。

浪漫主义的美学信条是风景如画。展望19世纪,有浪漫主义适合生存的沃土,那样的时代是由新能源和大机器主宰,社会和技术变革永久改变日常生活的范围和速度。英国的风景如画建筑始于虚假建筑,花园布局模拟中世纪建筑结构或样态。

随着哥特复兴的大规模兴起,浪漫主义趋向严肃。英国哥特复兴的领军人物奥古斯塔斯·普金认为哥特式是过去时代的道德和宗教价值的体现,而他自身所处的时代却匮乏的。1841年普金发表了《尖顶建筑或基督教建筑的真谛》,他列举了两条设计原则:"一是建筑中不应存在就便利性、结构和适宜性而言是多余的东西;二是所有装饰都只能是对建筑基本结构的美化。"他认为哥特是英国信仰、愿望和气候的唯一正确的表现形式,宣扬用它来建造包括现代世俗的产物,如火车站等的所有建筑。

(1) 浪漫主义风景。

威廉·肯特被视作英国园林景观传统的缔造者之一。景观建筑师不仅夸大而且"改进"了自然的品质。英国园林设计师不像法国人偏好严谨的几何格局,而是着力培育某种不规则的野趣。他们利用土地的自然形态,树木貌似自然生长,看似偶然为之,实则精心构思,营造宜人景观。同样地，连透过窗子欣赏到的景致都做了巧妙的设计，由近处层叠的树木，推及远处吃草的牛儿，凡目光所及均可如画。开凿凹陷沟渠阻止牛儿靠近房屋近旁的草坪，这样的奇思妙想让人惊叹不已！

兰斯洛·布朗是景观建筑风景如画态度的大力助推者。每当别人问他某地块的想法时，他总是说有"潜力"，因此人们就称他为"潜力布朗"。潜力布朗更倾向于一种新的、柔性的浪漫主义风格，模仿而非约束自然。在布伦海姆宫和查兹沃斯这样大规模的园林中，布朗营造出一番野趣，你见不到对称布局的花坛和笔直的步道，取而代之的是开阔的草坪，连绵的小丘上曲折的山路，点缀着自然种植的树木的溪流和池塘。布朗对英国风景的影响是无法估量的。他将全国上下的荒芜杂乱的农庄改造成为繁盛的庭园，而令英国至今为此而闻名于世。

英国造园家雷普顿（Humphry Repton）修正了这一风格，他认为最佳的布局应该是靠近房子的地方是规则的花坛，然后逐渐地过渡到周边自然的环境中去。由于受到 17 世纪的画家洛兰（Claude Lorrain）及其追随者的罗马乡村题材的画作的影响，这些所谓的英国式园林中会掺杂一些仿造的中世纪废墟和罗马的古庙。出现在像钱伯斯爵士（Sir William Chambers）这样的旅行家的游记中的中国园林的版画使得中国式的亭和其他异国风景也出现在英国园林之中。

英国的浪漫主义风格通过法国影响到了欧洲的其他各国，其中在法国一个比较著名的例子是在阿尔门维尔烈的园林。到了拿破仑统治的时期，工程师让·查尔斯·阿道夫采用浪漫主义风格建造了巴黎的许多大型园林，并立即产生了国际性的影响。在德国和奥地利，这种浪漫主义风格被庇乌克勒·穆斯考极力推崇，他在他柏林附近的地产上建造了极富浪漫色彩的穆斯考林苑，并且还在 1835 年出版了颇具影响力的《造园指南》一书。

托马斯·杰弗逊通过设计自己弗吉尼亚的蒙蒂塞洛宅邸，将浪漫主义风格引入北美。最著名的例子便是1857年由奥姆斯特德和沃克斯联手设计的纽约中央公园。这是在美国的第一个主要的公共的景观设计的例子，中央公园的成功在于它是市政建设与艺术成果的完美结合。1870年城市公园概念的形成也与中央公园有关，奥姆斯特德和沃克斯还设计了遍布全美的大部分城市公园。

（2）草莓山庄。

位于英国特维克娜姆的草莓山庄，1749~1777年由贺拉斯·沃波尔建造，是对哥特理念的拯救。贺拉斯·沃波尔也不管合适与否，把他喜欢的哥特母题自由组合在一起。想要完整地欣赏它的设计，必须从外面并且是三维角度来看。他的方法本质上而言极其浪漫主义，沃波尔想要草莓山庄呈现出天幕下丰富的轮廓，想要它在人们变化角度时，特征亦随之改变，想要它似乎在人们看来并非一次完工，而是随时间推移逐步成演进成这样子的。

草莓山庄建成时，正值先锋派的前沿领域关注"我们权威的英国传统"之时。英国哥特式前提在19世纪成为主题。

（3）奥古斯都·威尔比·诺斯摩尔·普金。

奥古斯都·威尔比·诺斯摩尔·普金是哥特风格的狂热的拥护者。他执着地认为中世纪的建筑优越于其他建筑时期，对19世纪英国及世界产生深远的影响。

普金的著作

普金认为中世纪是人类历史上最伟大的时期，那是人们为上帝的荣耀生活和工作，诸如大教堂一类的建筑也因此而得以建造。他憎恶文艺复兴时期的建筑，甚至包括18世纪和19世纪早期某些建筑师的哥特式建筑作品。他认为他们的作品在风格上不够正宗，用一些伪造的材料，比如用熟石膏来修补石质拱顶的填料。普金捍卫哥特式，源于哥特式不仅是"一种风格更是一种原则"。在普金看来，哥特式之所以是"真实的"，是因为它诚实地使用建筑材料，将结构暴露出来，功能由此而得以展示。由此，在后来理论家的眼中他成为功能主义的先驱。

1836年，普金出版了他的名作：《对比，或在14、15世纪高贵的大厦与现今类似建筑之间的比较》。这本书以图例对比了伟大的中世纪及其后的建筑用以表明当时趣味的衰败。通过此书，普金强烈地寻求中世纪天主教精神的纯洁，但是他所选择进行对比的一对建筑带有高度的偏见，中世纪的品质高，而较近年代的水准差。"对比"给普金带来声誉的同时也让他声名狼藉。

他后来出版的著作佐以插图表达了他对哥特式建筑、结构、设计及装饰的深入理解，观点更为客观公正，也更令人信服，达到了中世纪以来的最大成就。最有影响的有：《尖券或基督教建筑的真实原则》《基督教建筑现状》《基督教装饰及祭服汇编》。他的著作对整个维多利亚时期哥特式建筑的设计和建造产生了深远的影响，比他同时代所有人的贡献都大，甚至超越他本人的建筑作品，但这些作品常常饱受资金缺乏之苦。

普金的作品

普金设计了很多教堂，并负责扩建及改造若干修道院及大学建筑群。他把对哥特式建筑的认识和理解注入到这些作品中，尤其关注室内设计的细节，并喜用木材、金属、绘画及玻璃做装饰。他的部分作品后来被改建破坏，如萨瑟克大教堂，但大量作品仍然留存。主要的有伯明翰罗马天主教堂、德比圣玛丽亚教堂、斯塔福德郡圣伊莱斯教堂、圣奥古斯丁教堂、圣玛丽亚教堂，以及利物浦圣奥斯瓦德教堂。

此外，还留存了一些他改建或重新设计的房舍，如1845~1847年，他做了室内设计的兰开夏郡的一个巨大的带有又高又细角塔的哥特式房屋。这所房屋展示了某些细腻的哥特细节和雕刻装饰，直到建筑师去世后才完成。

威斯敏斯特宫

普金参与的最重要的建筑工程是新威斯敏斯特宫（国会大厦）。

1836年查尔斯·巴里在一次新宫殿设计竞赛中赢得了这一工程,竞赛明确要求新建宫殿必须是哥特式的,目的是强调英国议会制的自中世纪以来的历史延续性。由普金负责设计和监督所有内部墙体装饰、装饰画、彩色玻璃、日常器具、家居、地毯和所有装饰。普金的哥特细节如水平层拱平抑了典型的哥特式教堂的垂直性。整座宫殿是一座有1000个房间的巨大建筑体,沿泰晤士河的立面跨度900英尺(275米)。威斯敏斯特宫拥有经典的对称立面,外罩镶板的哥特外衣,如画的塔楼,包括著名的钟楼大本钟在内,创造出浪漫的、非对称的轮廓。宏大的规模、彩色玻璃及哥特式尖顶形式是为了创造出敬畏和仰慕的情绪,现在被用作议事和司法的场所。

尽管在第二次世界大战中受到损害,普金的很多成果还是得以幸存,显示出他的完美性和他的专业设计和技艺的学术品质。威斯敏斯特宫是普金最伟大的纪念碑。

2. 功能主义

一般来说,两次大战之间的时代特征是要寻求一个安全普遍的基础,也就是原则。"功能主义"这个词本身就显示了它的态度和目标。功能主义建筑师相信新的建筑师是智力和技术条件在现时代的不可避免的逻辑产物,并从中寻求基本原则。但是这种暗含的科学方法并不会阻止丰富而有意义的建筑作品的产生。寻求原则和处理实际任务如社会居所和市政环境相协调和相引导。

决定功能主义表达两个目的:形式的创建和本质涵义的恢复。通过运用独立的、规则的框架结构,自由平面可以实现第一个目的。为了实现第二个目的,倾向于基本的立体几何体量,避免传统母题和装饰。

功能主义思想在勒·柯布西耶的工程和理论中得以最典型的体现。勒·柯布西耶在《走向新建筑》(Towards a New Architecture,1923年),阐明当一个事物满足了某种需求,它就是美的。同时他还相信建筑本身就是能为人们带来快乐的事物。

基于对人和建筑的强烈信念,功能主义目标积极向上。功能主义者的核心信念是人类环境的质量可以通过重新获得真实而根本的意义的新建筑而得到提升。功能主义建筑的基本形式和严格原则可以解释成这样一种主张：反对建筑师们所谓历史主义的廉价母题和学究式句法，但首先所有的功能主义者都有一个基于对人类和建筑的强烈信仰之上的积极目标。由此，它获得了成为国际性运动的重要性和力量。

（1）勒·柯布西耶。

勒·柯布西耶的原则和卓见为现代人居城市发展做出的真正伟大贡献之一。

自由平面

在1914年柯布西耶就结合他的多姆－伊诺住宅提出了自由平面的基本概念。"我们已经创造了一种建筑的方法——框架结构——它完全独立于住宅平面的功能要求……允许室内格局的无数组合方式，以及立面上想象得到的任何光线的处理。"他由此认识到框架结构和开放空间之间的天然联系，从而使这些仅仅被少数先锋建筑师直觉感受到的东西变成普遍可能。

路德维希·密斯·凡德罗后来详尽阐述了这一概念，"自由平面是一种新的概念，它就像新的语言一样有着自己的语法。（比如说）一般的转角就给人以厚重的印象，这使得它很难与自由平面结合。许多人认为自由平面便意味着完全的解放，这是一种误解。自由平面和传统平面一样，要求建筑师有着很深的修养和理解。""自由平面和清晰的结构不能分离开来。结构是整体的支柱，它使自由平面成为可能。没有支柱的平面就不可能自由，而只能是混乱和凝滞。"

立体花园城市

柯布西耶并不认为城市即是由绿地、大的居住单元自由组合在一起构成的。他还提出了"立体花园城市"，一个自由空间环绕许多巨大居住结合体，它以一种新的形势体现了传统村庄的特点和个性，同时还恢复了那些"基本的快乐"，如阳光、空地和绿地等。勒·柯布西耶将城市的功能定义为"居住、工作、游憩和交通"。

交通分流

勒·柯布西耶还把市政中心作为一个城市要素引入，和其他要素一样，它成为居住功能的辅助性的延伸。其中非常重要的还有他的交通分流概念，这是第一次有人提出要将机动车交通和步行交通区分开来。它不仅是个实践性理念，其出发点是不同交通工具意味着不同的运动节奏，因而需要不同空间的认识。柯布西耶为人类居住区的三个基本类型给出了自己的一般性原则：农业生产单元、线性工业城镇和中心放射状的社会城市。所有这些他都看成是通过道路相联系的结构性场所。另外，他还对已有城市做了一系列有创见的规划，这些城市的原则和所处的环境都以不同方式相互作用着。其中为北非海岸的内穆尔所作的设计异常突出地体现了他的想法。

勒·柯布西耶的"新建筑五点"

勒·柯布西耶后来又将他的基本理念发展成了著名的"新建筑五点"，其中列举了新设计方法的具体优点，并由此定义出功能主义建筑的一般属性，简言之：即底层架空、屋顶花园、自由平面、带形长窗和自由立面。具体来说，即用细柱抬高建筑离开地面，让连续的绿地在建筑下面通过；由于城市中地面已经被建筑充满了，因此将把公园抛向天空作有效的屋顶花园；由大柱距的空间结构体系必然带来开敞式的平面布局，其中可以安装隔断来划分空间；矩形窗不受柱距开间尺寸的限制，采光面积更为有效；外墙不承重，自由开闭的幕墙和隔断满足功能与美观的需要。

勒·柯布西耶把这些原则运用到20世纪末期的主要作品中，从中确立了基本的建筑特征。这些作品包括：魏森霍夫住宅区、日内瓦国际联盟总部、普瓦希萨伏伊别墅、巴黎瑞士学生宿舍。此外，它还将原则的定义与类别的发展结合起来。在长期研究标准化住宅之后，他于1921年在希特罗汉住宅找到了突破点，该研究的最高成就是后来的马赛公寓，其中每三层配有一条走廊，为两个跃层的住宅单元服务。1929年首次出现的螺旋形博物馆体现了创立新类型的欲望。

萨伏伊别墅

萨伏伊别墅于1928~1931年在普瓦希建成至今，一直是功能主要建筑的典范。柯布西耶在这里实现了他的"新建筑五点"，并创造了一个有着无与伦比的力量和诗意的作品。在他的《作品全集1929~1934》一书中对该别墅的阐述体现出其设计目的和方式都是慎重考虑后的选择。

一般而言，设计取决于建筑师希望怎样结合结构的严谨性和空间的自由度，也取决于场地的特定条件和车行距离。为了获得更好的景观和隔绝地面的湿气，自然要将主要楼层架空于脚柱上。这样地面层就用来布置环绕门厅的车道及服务设施。汽车的单一运动轨迹决定了平面在大体上的对称。入口则配合车道的位置，设在弧形玻璃墙正中间的纵轴线上，显示出某种具有代表性的特征。然后就是处于建筑中央、将三层的空间联系起来的壮观的坡道。勒·柯布西耶在这里，为到达的功能赋予了一个令人信服的现代的解释，同时将竖向维度结合到自由平面之中。

到那时为止，开放空间还仅仅限于一般的通透或水平面上的自由运动。但在萨伏伊别墅，人们体验到的是三维空间的占有，也就是一种新的象征性的自由。然而这种体验与同样有着象征意义的科学技术的秩序同时发生，甚至可以说产生于这种秩序。对此勒·柯布西耶写道："在这座房子里我们经历的是真正建筑的漫步，呈现与眼前的是不断变化、出乎意料，甚至让人惊叹的景象。有趣的是，如此的丰富多彩变化都源自从设计角度出发的对梁柱的一丝不苟的安排。"通向主要楼层的坡道本身就是一个用空间来适应不同功能的巧妙试验，坡道继续向上，可以到达屋顶的"日光浴室"，在那里，曲线的墙体综合了随意、私密以及周围景观的相互作用。其中有一个特别重要的细节，是当参观者沿着坡道进日光浴室时正对的墙上设计的开口。这暗示着坡道并不仅仅通向屋顶、还通向开放的外部空间。这让关于路径的建筑主题得到了生动的体现。

萨伏伊别墅空间的丰富和动感完全被收容在一个近似方形的体量中，这实现了勒·柯布西耶的两个及本意图：对空间自由度和对基本形式的要求。长条形的连续洞口暗示着内部平面的自由，而总体的外表则体现出占典纯粹的特征。在《走向新建筑》一书中，他将立方体、圆锥体、球体、圆柱体和棱锥体看成是"伟大的基本形体"，并认为"建筑就是在光线下各种形体的巧妙、正确而杰出的组合。"所以，萨伏伊别墅，体现了对启蒙时期的人们对基本真理的探求。作为原型的抽象与理想化的体量源于普遍的自然属性，而它们也有着高度普遍的意义。萨伏伊别墅单纯的主要体量将住宅与宽广的自然和城市环境结合起来，而它的内部体量则实现了私人领域的拓扑关系。萨伏伊别墅的意义正在于这些特征的结合，正如文丘里指出的那样"它严格的、近乎正方体的外表包裹着一个复杂的内部结构，从开口和顶部的突起物我们可以窥之一二……它的内部秩序与住宅的繁复功能、居住的尺度以及出于私密性而产生的部分神秘感正相吻合。而其外部秩序则综合地表达了让住宅以一种容易协调的尺度与它所控制的绿地，以及有一天会融入其中的城市相协调的想法。"

（2）路德维希·密斯·凡德罗。

路德维希·密斯·凡德罗，作为另一位功能主义建筑的领军人物，行动上甚至比勒·柯布西耶更为注重类型和原则的发展。

表达的综合

路德维希·密斯·凡德罗建造了五个工程，每一工程都体现了一种"典型的"概念，并用尽可能经济的方法来进行设计。这些建筑普遍都试图表现出其美国原型的本质特征。前两个项目是摩天楼，第三个是连续楼板的水平向建筑，这让人想起沙利文的卡森—皮里—斯科特百货公司大厦。最后两个住宅很显然与赖特的十字形平面（后来密斯还研究过大厅的问题，并将庭院式住宅作为一种原创的类型加以发展）。在两个摩天楼的设计中，建筑简化到一个有轻巧的全玻璃幕墙包裹着的结构框架。1922年密斯对此做出的解释清楚的说明，他的出发点就是新技术，特别是钢和玻璃构成的框架结构技术："摩天大楼在建造的时候完全暴露了他们大胆的结构模式……然而，当外墙安装到位时，作为所有艺术美的基础的结构系统，便被一堆没有意义的、琐碎的混乱形式所掩盖了……如果我们用玻璃来代替外墙的话，我们就可以最清楚地考到清新的结构原则。在密斯两个居住项目中，则没有运用任何框架结构，而是将注意力集中在空间连续性的问题上。在砖砌住宅中，源自赖特的长长的、有方向性的墙体被断开、并置，从而形成一种无处停息的流动空间。在后来的设计中，密斯将自己的方法发展成空间表达性处理的一种系统语法。密斯最重要的一步，是巴塞罗那馆的设计，在这里，划分空间的屏障与给予自由平面一定秩序感的规则钢框架结合在一起，实现了19世纪出两大新事物的综合：具有开放重复性的秩序感的框架结构，一级弗兰克劳埃德赖特的既具流动性又具表达性的空间。勒柯布西耶曾预言过这种结合，密斯则最终得出了一个符合逻辑的结论。密斯自己也完全意识到他这一成就的重要性。

图根特哈特住宅

萨伏伊别墅在勒·柯布西耶全部作品中所表现的，在路德维希·密斯·凡德罗，则体现在布尔诺的图根特哈特住宅。密斯早期的另外两个杰作——巴塞罗那馆和1931年的柏林建筑博览会都已经得到了深入的理论研究，而图根特哈特住宅解决了一个混凝土的建筑任务，体现了密斯的设计准则是如何与住宅的复杂功能相适应的。

该住宅建在一个坡地上，入口位于建筑的顶层，顶层有卧室。一

第六部分　建筑文本翻译拓展训练（二）汉译英

部楼梯往下通向起居室，起居室通过露天平台和大台阶与更低处的花园相连。这一布局很复杂，更重要的是，它完全不同于一般印象中密斯的理想：一个置于水平基座上的简单的玻璃方盒子。因而图根特哈特住宅被认为带有一种"折中"的特性，但实际上这种判断是毫无根据的。反之，这个住宅应该说充满了空间表达性处理的丰富试验，它正体现了密斯的方法中包含的多种可能性。

从道路方向看，这个住宅是个低矮的水平向延伸的建筑。其顶层基本包括三个空间上独立的体量。一个是主人的卧室，一个是小孩的房间，还有一个是车库和司机的套房。这些体量被处理成方盒子，窗户则如同墙上的洞，体现出其中功能的相对封闭和私密的特性。然而它们以一种重叠的方式并置，预示着其间一个流动空间的存在。这些方盒子与巴塞罗那馆中独立墙体的空间功能相似，只不过墙体在这里变得更厚，而且中间被挖空了而已。在图根特哈特住宅中，方盒子之间的空间从入口一直通到平台。其中一半被覆盖于与方盒子相连的屋顶之下，这屋顶的一部分由钢柱支撑着，体现了主要层结构系统的连续性。人们必须通过一面弧形玻璃墙才能进入室内，它使得内外空间在功能交接处融合为一体。

主要层包括一个非常大的开放式起居空间，和一个比较封闭的厨房部分。起居部分通过条纹大理石的直墙面和望加锡乌木的弧墙进行表达性处理，他们分别划分出了起居室、餐厅、书房和大厅四个次一级的领域。书房位于空间中最深处，呈半封闭状态，而起居空和就餐空间则通过连续玻璃墙面向景观开放，这面玻璃墙只要按一个按钮就能完全滑入地下。建筑侧面是两重玻璃墙，两层之间是一个温室，可以为室内提供一整年的绿色。主要空间的封闭与分隔主要靠可移动的幕帘来实现。为了理解密斯的空间表达性处理，我们必须研究这些不同类型的墙如何交接在一起，因为所有的交集点和细部都有着组织空间的功能，并且体现着潜藏其中的开放空间的普遍概念。划分空间的元素还跟建筑的结构框架紧密相关，这一框架为主要楼层提供了规则的韵律感。十字形铬柱体现了整个系统的精确性和普遍的开放性。这正验证了密斯的名言，即一座"清晰"的建筑应该使得自由的平面具有可能，也就是说，具有意义。

在1931年为柏林建筑展览会设计的住宅里，密斯用清晰的结构和自由的平面结合创造出了一个令人信服的范例，而在他30年代所设计的法院建筑中，则是先对狭小的城市用地运用了一种大体上封闭的有机性的原则。战后他的晚期住宅作品（范思沃斯住宅，1946年以后）则体现了对结构表达性问题的进一步关注，但其中的基本概念还是一样的。

（3）沃尔特·格罗皮乌斯。

1911年，沃尔特·格罗皮乌斯设计了法古斯工厂，他把墙变成了钢和玻璃的轻质帘幕，厚重的转角消失了。而在他为德意志制造联盟的科隆展览会设计的"模范工厂"里，玻璃幕墙的连续性得到了强调，并与一个极富动感的元素——透明螺旋楼梯结合了起来。

包豪斯

包豪斯大楼如今还矗立在东德的德绍市内，其本身就有着特殊的涵义。自1919年以来，它成为了教授现代建筑和设计的机构，也成为勾画新的生活场景的地方。在创始人沃尔特·格罗皮乌斯的领导下，产生了无法估量的影响。包豪斯简直成为了"现代设计"的代名词，而这个名字至今还保留着一些古老的神秘感。

包豪斯最初在魏玛成立，但1926年，它被迁入德绍由格罗皮乌斯自己设计的一幢更大些的新建筑里。在这里，格罗皮乌斯不仅为他的学校活动创造了一个实践的框架，同时还试图建立人类建筑环境视觉表现的新图景。"

在功能上，包豪斯包含有三个部分：教学区、工作室和学生宿舍。其中每一部分都占据单独的一翼，整个平面表现了对功能分区的典型愿望，同时实现了建筑与环境的积极联系。包豪斯的开放和动态的形式由三个主要体块间的连接元素进一步加强。学校和工作室之间通过跨越道路的天桥相连,天桥上设置管理部门和建筑系。而工作室和学生宿舍之间的联系则是一座较低的侧楼,里面是会议厅和餐厅。这样做的结果正吻合格罗皮乌斯的论断:"虚假的轴对称让位给了用自由而不对称的组群造成的生动而富有韵律的均衡。"建筑外立面的变化表现出内部空间类型的不同,教室和行政管理部分有连续的带形窗,而工作室部分则完全由玻璃幕墙包围,学生公寓则用突出的阳台划分开口。每一片墙件都像薄膜一样包裹在钢筋混凝土的框架之外。一种透明和反射的神奇效果由于工作室的大玻璃墙营造了出来。在晚上,整个建筑从内部被照亮,变成了一个大而透明的光的容器,展现着开放而清晰的空间的动态感。

包豪斯的教育哲学是要寻求一种艺术与技术的新的结合,也就是"赋予机械产品真实而有意义的内涵"。为了做到这一点,包豪斯还着眼于开拓个人的自我表现能力并提高人们基于科学知识的客观审美能力。这一理念将包豪斯置于传统和启蒙运动的精神之中,而它的巨大成功正是在于它体现了那个时代的功能主义精神。从1926~1928年格罗皮乌斯离开学校,两年之中包豪斯取得了其他建筑20年才能取得的成就,要了解包豪斯以及更普遍的功能主义建筑,就要像掌握它对于秩序的理想一样,同时理解它的自由观念。格罗皮乌斯将建筑师称为"协调者"而不是一个传统意义上的艺术家,因为自我表达并不等同于自满和专制的任意妄为,而应是创造性地运用科学探索和技术开发成果的能力。秩序首先意味着建立一种标准,能够用来容纳富有创造性的多样化。

完全剧场

功能主义建筑的人性目标和潜在的丰富性很好地体现在了当时最迷人的作品之一,格罗皮乌斯于1927年设计的"完全剧场"。在整个剧场中,演出者和观赏者不再像文艺复兴时期那样彼此隔绝,格罗皮乌斯希望能"将观赏者置于戏剧活动的中心"。为了实现这一目的,他将一部分看台设计成可活动的,这样,舞台不仅可以被观众环绕,根据他的需要,还可以通过墙上的胶片投影来"用光线进行建造"。其中一个更根本的创新在于引入了一个围绕全场的"环形舞台"。通过打开四周的墙面,观赏者可以体会到置身于演出中央的感觉。在整体效果上,"完全剧场"正是象征性地体现了作为变化不息的"完全"动态世界的一分子的人类的新的存在环境。换句话说,它赋予了人们开放空间概念的内容和意义。

(4) 魏森霍夫住宅区。

1927年,德意志制造联盟在斯图加特举办了一个名为"居住"的展览。这个展览希望用一组件住宅来展示现代主义运动所构想的新的生活环境。路德维希·密斯·凡德罗,作为制造联盟的副主席,被选举作该项目的理事。他首先提出一个迷人的规划(1926年),将地中海村庄一般的多样性和现代建筑的流动空间概念结合在一起。在展览结束后需要把这些住宅卖掉时,其设计做了些后期改动以便让每个单体的独立性更强。密斯邀请了许多现代主义运动的领袖人物,如勒·柯布西耶、格罗皮乌斯、奥德、斯塔姆、陶特兄弟、夏隆、弗兰克,拉丁,德克尔、海尔贝斯门、布尔热瓦、施里克等参与其中,因此这次展览真正成为了一种新风格的展示。菲利普·约翰逊在1947年恰当地将这个住宅区称之为"现代建筑史上最重要的一个建筑群"。虽然魏森霍夫住宅区在战争中遭到了很大的破坏,但它的总体环境特征被保留了下来,除了格罗皮乌斯的住宅外,那些最有趣的建筑至今还在,如密斯、勒·柯布西耶、奥德、斯塔姆和夏隆设计的住宅。

今天,功能主义住宅运动常常被批评是枯燥的、忽视环境质量的。因此魏森霍夫住宅区里可以体验到的多种多样的舒适感和人体尺度,多少让人有些惊讶。尽管这里有着使用共同的形式语言的明确要求,但建筑师的不同个性还是十分明显:从密斯精确的优雅,到奥德和斯塔姆克制的简约,以及夏隆的动态体量。由此可以明显的看出,功能主义者实际上并没有那样忽视环境质量，反而是今天可能更加缺乏这些东西。

住宅区里小单元形成的多样化的小环境是在一个主体建筑的背景之下，那就是密斯·凡德罗设计的公寓。虽然这座建筑看上去是个统一的体量，但其规则的钢框架使室内新的开创性自由设计成为可能。12个公寓没有两个是相似的，它们全部都用非承重隔墙来满足不同的需求。就像杰迪昂在描述密斯的建筑时指出的那样，其目的是为给予使用者根据自己的需求来细分空间的可能性，而"工厂必须创造出技术上能满足这种需求的墙体"。勒·柯布西耶的两个住宅中较大的那个引入了一种更加开放的设计类型。它大体上依照"新建筑五点"来建造。但主要层被设计成一个连续空间，这样便可以随意安排用于睡觉的浅凹室空间。勒·柯布西耶这样解释他的设计："只有晚上才用的大房间，用可移动的隔墙来分隔，这样便把房间变成了一辆卧车……一条与国际卧铺列车公司的经济舱一样宽的小走廊，在晚上就如同一截独立的车厢。"只有书房是与公共空间隔开的，它被放在与屋顶平台等高的平面上。勒·柯布西耶设计的另一个住宅的设计则基于西特罗翰住宅的模式。它有一个两层通高的起居室，预示了勒·柯布西耶在马赛公寓的典型居住单元的做法。

由此可以看出密斯和勒·柯布西耶都借助建魏森霍夫展览表达了他们的一些基本观点，格罗皮乌斯也不例外。他设计了一个有着技术独创性的住宅，它由一个轻巧的布置紧凑的钢框架支撑起厚软木的保温墙，室外和室内分别铺设了石棉板和隔音板。这里引入的干挂法是对建筑工业的一个重要贡献。

从城市的总体规划，以及一种现代生活形式的居住设计，到技术和经济问题的解决，魏森霍夫都体现了一种伟大的集体成就，它很快就成为了一个建筑学的朝圣地。

6.2 参考译文

Romantic and Functional Architectures

1. Romanticism

With its emphasis on untamed nature, the picturesque, the past, and the exotic, Romanticism was a reaction against order and regularity. Romantics delighted in the asymmetrical and irregular for their highly picturesque qualities. Ruins in general became a Romantic obsession, as their ruggedness, wildness, and fragmentation illustrated the powerlessness of men and women in the face of irresistible natural forces and the melancholy relentless erosion of their works over time.

The esthetic doctrine of Romanticism was the Picturesque. Looking forward into the nineteenth century, it is equally clear that romanticism could found fertile ground in an era that would become dominated by new sources of energy and huge machines, as the fruits, both sweet and bitter, of rapid social and technological change altered forever the scale and speed of daily life. Picturesque architecture in England began with follies, the playful use of medieval – inspired structures or ruins of structures as vocal points in the layout of gardens.

Romanticism acquired a more serious tone with the widespread emergence of the Gothic Revival. In England the leading Gothic Revivalist was Augustus Welby Northmore Pugin (1812 – 1852), who regarded the Gothic as an embodiment of proper moral and religious values from the past that he felt were all but absent in his own time. In 1841 Pugin published The True Principles of Pointed or Christian Architecture, in which he enumerated his ideals: "First, that there should be no features about a building which are not necessary for convenience, construction, and propriety, and second, that all ornament should consist of the enrichment of the essential construction of the building." He regarded the Gothic as the "only correct expression of the faith, wants and climate" of England and advocated its use for all buildings, including modern secular ones such as railroad stations.

(1) The Romantic Landscape。

William Kent is regarded as one of the founders of the English landscape-garden tradition, in which the landscape architect exaggerated and "improved" on natural qualities. Instead of the rigid geometric plantings favored by the French, English garden designers cultivated a certain irregular wildness. They exploited the natural contours of the land, formed trees into apparently natural pattern, and developed seemingly fortuitous, but actually carefully contrived, views of carefully sited buildings. Likewise, equally carefully planned views from the windows of buildings extended beyond nearby clumps of trees into the more distant landscape, where cows grazed as in a landscape painting. A ditch with a fence or hedge at its bottom prevented the cows from encroaching upon the lawns in the immediate vicinity of the house; discovery of this concealed barrier was a surprise, causing one to laugh or exclaim.

Lancelot Brown (1716 – 1783) was the leading promoter of this picturesque attitude toward landscape architecture. When he was asked about his opinion of any piece of ground, he would say it had "capabilities" and thus he became known as Capability Brown. He preferred a new, softer romantic style that imitated rather than disciplined nature. At such great houses as Blenheim Palace and Chatsworth, Brown replaced the parterres of symmetrically arranged flower beds and straight walks with sweeping lawns, sloping hills with curving paths, and rivers and ponds punctuated by informally planted groups of trees and shrubbery, to achieve the effect of a wilderness. Brown's impact on the English landscape can hardly be overstated. Working throughout the length and breadth of the nation, he transformed large areas of unkempt countryside into the kind of lush parkland for which England has become renowned.

The English landscape architect Humphry Repton modified the style, believing that a house was best set off by formal flowerbeds that merged by subtle degrees into a naturalistic background. These so – called English gardens often incorporated "follies" fake medieval ruins and Roman temples, inspired by the 17th – century paintings of the Roman countryside by Claude Lorrain and his followers. Chinese pavilions and other exotics were inspired by engravings of Chinese gardens in travel books by such travelers as the architect Sir William Chambers.

The English romantic style spread to the rest of Europe by way of France, where a notable example of the style was created at Ermenonville. As late as the reign of Napoleon the English romantic style was employed by the engineer Jean Charles Adolphe Alphandin laying out the great parks of Paris, which exerted an immediate international influence. In Germany and Austria, the romantic style was enthusiastically endorsed by Prince Hermann von Puckler – Muskau, who created a romantic park on his estate near Berlin and published the influential Hints on Landscape Gardening (1835).

The romantic style was introduced in North America by Thomas Jefferson at Monticello, his Virginia estate. The most important example of this style is Central Park, New York City, designed in 1857 by Frederick Law Olmsted and Calvert Vaux. It was the first major public example of landscape architecture in the U. S. and was so successful both as a municipal enterprise and as a work of art that by 1870 it had influenced the creation of public parks, many of them designed by Olmsted and Vaux, all over the nation.

(2) Strawberry Hill。

Very free in its use of gothic motifs in combinations, Strawberry Hill, Twickenham, England, a country estate, built by Horace Walpole between 1749 and 1777, served as an approach to the rescue of gothic ideas. Walpole selected any motif he liked, combining them without regard for scale or appropriateness. To appreciate the house's design fully, one must view it externally and in three dimensions. His approach being essentially romantic, Walpole intended for Strawberry Hill to present a rich silhouette against the sky, intended that its character should change as one moved around the perimeter, and intended for the ensemble to appear as though it had been built not all at one time, but had grown up randomly over time as had its medieval inspirations.

When Strawberry Hill was built, a new interest in "our authentic English past" was the very cutting edge of the avant – garde. This premise of gothic as English was to continue into the nineteen century as a major theme.

(3) Augustus Welby Northmore Pugin (1812 – 1852)。

Augustus Welby Northmore Pugin was a fanatical enthusiast of Gothic. His passionate belief in the superiority of medieval over all other architectural periods was to have a profound impact on nineteenth – century architecture in England and beyond.

Augustus Welby Northmore Pugin's Publications

He regarded the Middle Ages as the greatest era in human history, a time when men were inspired to live and work for the glory of God and when buildings such as the great cathedrals were erected in this spirit. He abhorred Renaissance architecture and even more vigorously the eighteenth – and early nineteenth – century Gothic work of some architects. He complained that not only was their work not authentic in style but that they employed sham materials as in, for example, using plaster to repair stone vaults or iron as a supporting medium. Pugin's defence of his chosen style, on the grounds that it was "not a style but a principle", involved the claim that Gothic was "true" because it was the result of an honest use of materials in which structure was exposed and function thereby demonstrated. Consequently Pugin was considered as forefather of functionalism in the eyes of later theorists.

In 1836 he then published his famous book which he called Contrasts; or a Parallel between the Noble edifices of the Fourteenth and Fifteenth Centuries and similar Buildings of the present day. Showing the Decay of Taste, the book was finely illustrated and compared the buildings of the great medieval past with those of later ages. It was a passionate plea for the purity of the medieval Catholic spirit but Pugin's choice of pairs of buildings to contrast was often highly prejudiced, the medieval example being of high quality and the more recent one of inferior standard. "Contrasts" brought him fame and notoriety.

Pugin's later published books evidencing more balanced and cogent arguments and illustrating his deep understanding of Gothic architecture, its structure, design and decoration. Most influential of his publications were The True Principles of Pointed or Christian Architecture. The Preset State of Ecclesiastical Architecture and Glossary of Ecclesiastical Ornament and Costume, which showed a comprehension greater than anyone else had achieved since the end of the Middle Ages. His books exerted a lasting and wide – ranging influence on the Gothic architecture designed and built throughout the whole Victorian period, greater than almost any other architect's contribution at this time and certainly greater than Pugin's own architectural work, commissions which often suffered from lack of adequate financing.

Pugin's Works

Pugin designed a number of churches and was responsible for enlarging and altering several monastic and collegiate building groups. In all of these he put into practice his knowledge and understanding of Gothic structure and he was especially interested in the detail of furnishings and decoration in wood, metal, painting and glass. Some of his work has been later altered or damaged as at Southwark Cathedral but a quantity remains. Notable examples include: the Roman Catholic Cathedral (St Chad) of Birmingham and the churches of St Marie, Derby, St Giles, Cheadle, Staffordshire, St Augustine, Ramsgate, St Marie, Uttoxeter and St. Oswald, Liverpool.

His work survives at a number of houses where he altered or redesigned part of the building as, for example, at Chirk Castle in Clwyd where he worked in the interior 1845 – 1847. The chief example is Scarisbrick Hall, Lancashire, a large Gothic house with a tall, and slender corner tower. The house displays some fine Gothic detail and carved ornament. It was completed after the architect's death.

Palace of Westminster

The most important commission with which Pugin was involved was the building of the new Palace of Westminster (Houses of Parliament). In 1836 Sir Charles Barry won the prestigious commission in a competition whose term demanded that the new building should be Gothic in an attempt to stress the continuity of the British parliamentary system which had started in Medieval times. Pugin made the designs and supervised the work for all the interior wall decoration, the decorative art, the stained glass, the fittings, furniture, carpets and all ornament. Pugin's Gothic details suppressed typical Gothic Cathedral verticality with horizontal stringcourses. The Palace was an enormous complex with 1000 room and a facade spanning 900 feet (275m) along the Thames River, classical in the symmetry of its river facade but clothed in panelled Gothic dress and with picturesque massing of towers, including the famous clock tower Big Ben, creating a romantic, asymmetric skyline. The grand scale, stained glass and pointed forms of the Gothic Cathedral were reused to create feelings of awe and admiration but now for a building devoted to government and law making.

Despite damage during the Second World War, much of Pugin's work survives to show his perfectionism and the scholarly quality of his professional design and craftsmanship. The Palace of Westminster is Pugin's greatest memorial.

2. Functionalism

On balance the period between the two world wars was characterized by a search for a secure common basis, that is, for principles. The word "Functionalism" is indicative of its attitudes and aims in architecture. Functionalistic architects sought basic principles in the belief that the new architecture was the inevitable logical product of the intellectual and technical conditions of the age. But the implied scientific approach did not impede the creation of rich and meaningful works of architecture. The search for principles was coordinated and directed towards dealing with practical tasks such as the social dwelling and the urban environment.

Two main endeavours determined Functionalist articulation: the establishment at a unity of form and function and the recovery of essential meanings. The first aim was satisfied by the free plan, which was made possible by use of the independent, regular skeleton structure. The second aim led to a preference for elementary stereometric volumes, and the avoidance of traditional motifs and ornament.

Functionalism is most typically represented in the projects and theories of Le Corbusier. In Towards a New Architecture (1923) Le Corbusier stated that when a thing responds to a need, it is beautiful and he also believed that architecture is a thing, which in itself produces happy people.

Functionalism had a positive aim based on a strong belief in man and in architecture. The very core of the Functionalist creed is that man's condition may be improved through a new architecture which reconquers true and fundamental meanings. The elementary forms and strict principles of Functionalist architecture may be interpreted as a protest against what the architecture considered to be the devalued motifs and academic compositions of historicism, and it possessed the significance and power to become an international movement.

(1) Le Corbusier。

Le Corbusier's principles and visions are one of the truly great contributions to the development of a modern and human city.

Free Plan

In 1914 Le Corbusier defined the basic concept of the free plan in connection with his Dom-ino houses. "We have then produced a way of building—a bone structure—which is completely independent of the functional demands of the house plan…allowing for numerous combinations of internal disposition and every imaginable handling of light on the façade." He thus recognized the natural relationship between skeleton construction and open space and made generally possible something that so far had only been intuited by a few pioneers.

Later Ludwig Mies van der Rohe elaborated the definition: "The free plan is a new conception and has its own grammar, like a new language. A normal corner (for instance) gives an impression of massiveness which it is difficult to combine with the free plan. Many believe that the free plan means absolute liberty. That is a misunderstanding. The free plan asks for just as much discipline and understanding from the architect as the conventional plan. " "The free plan and a clear construction can not be kept apart. The structure is the backbone of the whole and makes the free plan possible. Without that backbone the plan would not be free, but chaotic and therefore constipated. "

Vertical Garden City

Le Corbusier did not believe that large dwelling units freely placed in a green space of themselves constitute a city. He introduced the "vertical garden city," that is, a very large unite d'habitation surrounded by free space, which in a new form preserved some of the properties and the identity of the traditional village were preserved and at the same time "the essential joys": sunlight, space and greenness were restored. He defined the basic urban functions as "living, working, cultivation of the body and the mind, and circulation".

Differentiated Traffic

Le Corbusier introduced the civic centre as an urban element, as well as other minor extensions of the dwelling. Of particular importance is his idea of differentiated traffic, which first of all implies the separation of motorized and pedestrian communication. The idea was not merely a practical one but took as its point of departure the recognition that different velocities imply different rhythms, and hence require different kinds of space. Le Corbusier applied his general principles to three basic types of human settlement: the "unit of agricultural production," the "linear industrial town" and the "radio-concentric social city." He conceived all of these as structured places connected by routes. In addition, he worked out visionary plans for a series of existing cities, where the principles and the local circumstances interact in various ways. A particularly clear illustration of his ideas is offered by the project for Nemours on the North African coast

(1934).

Le Corbusier's 5 points

Later he developed the general concept into his famous 5 points d'une architecture nouvelle "Five Points Toward a New Architecture". In these he listed the concrete advantages of the new approach, and thereby defined the general properties of Functionalist building.

(1) Stilts raise the building above the ground to allow for spatial continuity and tree circulation;

(2) A roof garden gives us back the ground lost under the building and unites it with the surrounding landscape;

(3) A free plan makes the storeys independent of each other and allows for a meaningful and economic use of space;

(4) Continuous windows make the spaces open and in contact with nature outside;

(5) The free facade transforms the massive wall into a screen which may be opened or closed at will.

Le Corbusier applied these principles in his main works of the late twenties, where they determine the general architectural character: the Weissenhofsiedlung, Palais des Natrons, Geneva, Villa Savoye, Poissy and the Pavilion Suisse, Paris. Furthermore, he combined the definition of principles with the development of types. His Maison Citrohan from 1921 formed the point of departure for a long research on the standard dwelling, which culminated with the unite d'habitation in Marseilles, where two double height apartments are served by one corridor for every three floors. The desire for types is also illustrated by the spiral-shaped museum, a concept that appeared for the first time in 1929.

Villa Savoye

Since it was built in 1928-1931, the Villa Savoye in Poissy has been a classic example of Functionalist architecture. Le Corbusier illustrated here his 5 points and created a work of incomparable strength end poetic quality. The text that accompanies the presentation of the villa in his Oeuvre Complete 1929 – 1934 shows that his ends and means were a deliberate choice.

In general, the design was determined by the architect's wish to combine structural rigour and spatial freedom, and by the particular conditions of the site, as well as access by car. To obtain a better view and to avoid the humidity of the ground, it was natural to raise the main floor on stilts. Thereby the ground floor could be utilized as a driveway around the entrance hall and the services. The simple movement of the car determined the general symmetry of the plan. The entrance is placed in the longitudinal axis in the middle of the curved glass wall which accompanies the driveway, indicating a certain representative character. This prepares for the splendid ramp which at the very centre of the building rises up to connect the three levels. Le Corbusier gives here a convincing modern interpretation of the function of arrival, and at the same time integrates the vertical dimension in the free plan?

So far open space had been concretized as general transparency, or as free movement on the horizontal plane. In the Villa Savoye what is experienced is the occupation of three-dimensional space, that is, a new symbolic freedom. But this experience happens within the equally symbolic order of science and technology; we might say because of this order. Thus, Le Corbusier wrote: "In this house we are presented with a real architectural promenade, offering prospects which are constantly changing and unexpected, even astonishing. It is interesting that so much variety has been obtained when from a design point of view a rigorous scheme of pillars and beams has been adopted." The ramp leads up to the main floor which is a masterly exercise in spatial adaptation to various functions, and from there on to the "solarium" on the roof, where curved walls represent a synthesis of informality, privacy and interaction with the surrounding landscape. A particularly significant detail is an opening in the wall which faces the visitor when entering the solarium. This means that the ramp not only leads to the roof, but into open space. It makes the architectural theme of the path a living reality.

The spatial richness and dynamism of the Villa Savoye is contained within an approximately square volume, which allows Le Corbusier to satisfy both of his basic intentions: the desire for spatial freedom and the desire for elementary form. Long, continuous openings indicate the free plan within, but the general exterior appearance is characterized by classical purity. In Towards a New Architecture he called cube, cone, sphere cylinder and pyramid "the great primary forms" and said that "architecture is the masterly, correct and magnificent pay of volumes brought together in light." Villa Savoye thus represents the Wish of enlightened man for a return to elementary truth. As archetypal abstractions the Platonic volumes refer to general natural properties, and their meaning is highly public. The simple main volume of the Villa Savoye serves to integrate the house in a wider natural and urban context, whereas the interior concretizes the topological relationships of the private domain. The meaning of the villa consists in the combination of these characters, as has been pointed out by Venturi: "Its severe, almost square exterior surrounds an intricate interior configuration glimpsed through openings and from protrusions above. ... Its inside order accommodates the multiple functions of a house, domestic scale, and partial mystery inherent in a sense of privacy. Its outside order expresses the unity of the idea of house at an easy scale appropriate to the green field it dominated and possibly to the city it will one day be part of".

(2) Ludwig Mies van der Rohe。

Ludwig Mies van der Rohe, the other leading exponent of Functionalist architecture, even more than Le Corbusier centred his activity on the development of types and principles.

Synthesis of Articulation

Between 1919 and 1924 Ludwig Mies van der Rohe produced five projects for office buildings and one – family houses. Each project represents a "typical" concept, and is worked out with the greatest possible economy of means. In general they attempt to define the essential properties of American prototypes. The two first projects are skyscrapers, in which the building is reduced to a structural skeleton covered by a light curtain wall entirely of glass. He believed the new structural principles were clearly observed when glass was used instead of the outer walls. The third a horizontally directed building with continuous floor areas, reminiscent of Sullivan's Carson – Pirie – Scott store. In the two last ones are houses no skeleton is used, but instead attention is centred on he problem of spatial continuity.

In later works Mies developed his method into a systematic grammar of spatial articulation. The decisive step was taken with the Barcelona Pavilion (1929) where space – defining screens are combined with a regular steel skeleton which gives order to the free plan, achieving a synthesis of the two main innovations of the nineteenth century: the open, repetitive order of skeleton construction and the fluid, but articulate space of Frank Lloyd Wright. This synthesis was anticipated by Le Corbusier, and brought to its logical conclusion by Ludwig Mies van der Rohe, Mies was fully aware of the importance of his achievement.

Tugendhat House

What the Villa Savoye represents in the oeuvre of Le Corbusier, the Tugendhat House in Brno represents for Ludwig Mies van der Rohe (1886 ~

1969). Whereas the two other early masterpieces by Mies, the Barcelona Pavilion and the House at the Berlin Building Exhibition in 1931, were theoretical studies, the Tugendhat House is the solution of a concrete building task, and shows how Mies's principles may be accommodated to the complex functions of a dwelling.

The house was built on a sloping site, with its entrance on the level of the upper floor, which contains the bedrooms. A staircase leads down to the living room, which is connected, with the garden below by means of a terrace and a wide flight of stairs. The layout is complex and in important respects differs from what is supposed to be Mies's ideal: a simple rectangular glass box placed on a horizontal podium. The Tugendhat House has therefore been characterized as a "compromise", but this judgment is unacceptable. Instead the house should be considered a particularly rich exercise in spatial articulation, which illustrates the possibilities inherent in Mies's method.

From the street the house appears as a low, horizontally extended building. Basically the upper floor consists of three spatially separate volumes. One contains the owners' bedrooms, one the children's rooms, and one the garage and chauffeur's suite. The volumes are treated as boxes, with windows placed like holes in the walls, expressing the relatively enclosed and private character of the functions they contain. They are juxtaposed however, in such a way that they overlap and indicate a fluid space between them. The boxes therefore have a spatial function analogous to that of the freestanding wells in the Barcelona Pavilion, only here the walls have become thick and hollow. In the Tugendhat House the space between the boxes leads from the entrance to a terrace. It is half covered by a roof connecting the boxes, and partially resting on steel columns, which represent a continuation of the structural system of the main floor. To enter the house, one has to go along a curved glass wall which causes outside and inside space to fuse at the point where they are functionally connected.

The main floor consists of a very large openly planned area for living, and a more enclosed kitchen section. The living area is articulated by means of a straight wall of onyx and a curved wall of Macassar ebony which define four subordinate domains: living room, dining room, study and hall. The study is located in the inmost point of the area, and has the character of a semienclosed pocket, but the living and dining areas open up on the landscape through continuous glass walls, which at the press of a button could slide into the floor. Along the side is the double glass wall, and the intermediate space contains a conservatory which provides greenery the whole year round. The enclosure and sub-division of the main area could be regulated by means of movable curtains. To understand Mies's spatial articulation, it is important to study how the different types of walls are joined together, for all joints and details are a function of the spatial composition and express the general underlying concept of open space. The space-defining elements are also carefully related to the structural skeleton which forms a regular rhythm throughout the main floor. Cross-shaped chromium columns express the precision and general openness of the system. They thus prove Mies's dictum that a "clear" structure makes the free plan possible, that is, meaningful.

In the house for the Berlin Building Exhibition in 1931, Mies gave a particularly convincing demonstration of the combination of clear structure and free plan, and in his courthouses from the thirties applied the principle to generally enclosed organisms on relatively small urban building sites. His later houses from the postwar period show an increased preoccupation with the problem of articulated structure, but the basic concept remains the same.

(3) Walter Gropius。

In 1914 Walter Gropius transformed the wall in the Fagus factory into a light curtain of iron and glass, and the massive corner is eliminated. In Gropius's "model factory" for the Werkbund exhibition in Cologne (1914), the continuity of the glass curtain was emphasized, and incorporated the dynamic element of a transparent spiral staircase.

Bauhaus

The Bauhaus building is still standing in Dessau in East Germany. The Bauhaus had a particular meaning. Since 1919 it had been the institution where modern architecture and design were taught and where the image of a new way of life was being shaped. Under its director, Walter Gropius (1887~1970), it exercised an influence that can hardly be overestimated. "Bauhaus" simply became synonymous with "modern design" and the name still retains some of its old mystique.

The Bauhaus started its activities in Weimar, but in 1926 it was moved to a new and larger building in Dessau, designed by Gropius himself. Here Gropius not only created a practical frame for the activities of his school, but also sought to make the new image of man's architectural environment visually manifest.

Functionally, the Bauhaus consisted of three parts: the school proper, the workshops and the students' dormitory. Each of these was housed in its own wing, a plan which demonstrated a typical wish for functional differentiation and at the same time created an active relationship between the building and its surroundings. The open and dynamic form of the Bauhaus was furthermore emphasized by connecting elements between the main wings. The school and the workshops were joined by means of a bridge spanning across the street of access, and this bridge contained the administration and the architectural department. The workshops were joined to the students' building by a lower wing containing the meeting hall and the dining room. As a result the building was in line with Gropius's statement: The hollow sham of axial symmetry is giving place to the vital rhythmic equilibrium of free asymmetrical grouping. The exterior wall treatment was varied to express the type of space behind, so the classroom wing and the administration had continuous window bands, the workshops an enveloping curtain wall of glass, and the students' building separate openings with projecting balconies. Everywhere the wall appeared as a thin membrane wrapped around the reinforced concrete skeleton. A fascinating effect of transparency and reflection was created by the great glass wall of the workshops. At night the whole building, lit from the inside, became a sort of large transparent light-modulator which expressed the dynamism of open, but articulated space.

The educational philosophy of the Bauhaus sought a new synthesis of art and technology that is, "to give the products of the machine a content of reality and significance." To achieve this the Bauhaus simultaneously wanted to free the individual's power of self-expression and to develop an objective aesthetics, based on scientific knowledge. This idea places the Bauhaus in the tradition and spirit of the Enlightenment, and its singular success is due to the fact that it represented the fundamental spirit of the age. In two years, from 1926 until 1928, when Gropius left the school, the Bauhaus achieved more than other schools of architecture have done in two decades. To understand the Bauhaus and Functionalist architecture in general, it is necessary to grasp its idea of freedom as well as its ideal of order. Gropius called an architect a coordinator," rather than an artist in the traditional sense as self-expression did not mean personal content and arbitrary caprice, but the power to use creatively the results of scientific investigation and technological development Order meant primarily the establishment of standards capable of creative variation.

Totaltheater

The human aims and potential richness of Functionalist architecture may be illustrated by one of the most fascinating projects of the period, Gropius's Totaltheater of 1927. In the total theatre actor and spectator are no longer separated as they had been since the Renaissance. Gropius's aim was to "pull the spectator into the very centre of the dramatic action." He achieved this by making a part of the auditorium mobile, so that the stage could be surrounded by the spectators and by his intention to use film projections on the wall surfaces to "build with light". A more radical innovation yet was the introduction of a "ring stage" engirdling the whole space. By opening the surrounding walls the spectators may experience being placed in the middle of the action. In its overall effect the Tetal theater is a symbolic expression of man's new existential situation as a participant in a "total" dynamic world of incessantly self - changing energies. In other words it gives human content and meaning to the concept of open space.

(4) Weissenhofsiedlung。

In 1927 the German Werkbund organized an exhibition in Stuttgart called The Dwelling. This exhibition was conceived as a group of houses which should visualize the new environment imagined by the modern movement. The vice president of the Werkbund, Ludwig Mies van der Rohe, was chosen as general director of the project. He first produced a fascinating plan (1926) which combined the intimate variety of a Mediterranean village with the fluid space of modern architecture. As the houses had to be sold after the close of the exhibition, the design was later modified to make each element more independent. Mies invited many of the leading exponents of the modern movement to participate: Le Corbusier, Gropius, Oud,

Stam, the brothers Taut, Scharoun, Frank, Rading, Docker, Hilberseimer, Bourgeois, Schneck as well as the pioneers Behrens and Poelzig, so the exhibition really became the manifestation of a new style and Philip Johnson in 1947 rightly called the Siedung "the most important group of buildings in the history of modern architecture." The Weissenhofsiedlung was considerably damaged during the war, but the general environmental character has been preserved, and except for Gropius's houses, the most interesting buildings are still standing, that is, those by Mies, Le Corbusier, Oud, Stam and Scharoun.

Functionalist housing developments are today often criticized for being sterile and devoid of true environmental quality. It is therefore somewhat surprising to experience the varied intimacy and human scale of the Weissenh of siedlung. Although there is a noticeable wish for a common formal language, the architectural character varies considerably from the calculated elegance of Mies to the restrained simplicity of Oud and Stam, and the dynamic volumes of Scharoun. From this it is apparent that the Functionalist approach did not exclude the environmental quality as such, which today is so much lacking.

The smaller units of the Siedlung form a varied milieu, given its backbone by a large apartment building by Mies van der Robe. Although it appears as a unified volume, the construction is a regular steel skeleton which allowed for a new revolutionary freedom of interior planning. None of the twelve apartments is alike: all of them were adapted to different needs by means of secondary partitions. The intention was "to give the user the possibility of subdividing his space according to his needs", and "the industry has to produce technically satisfactory walls for this purpose", as pointed out by Giedion in a description of Mies's building. A still more open type of planning was introduced in the large; of Le Corbusier s two houses. In general, the design follows his 5 points, but the main storey is here conceived as a continuous space, where shallow alcoves for sleeping may be created at will. Le Corbusier explains the design as follows: "The large room is made by dispensing with the movable partitions which are only used at night so as to turn the house into a sort of sleeping-car. …A little corridor the exact width of those in the coaches of the Compagie Internationale des Wagons-Lits acts as a passage solely at night." Only the study is separated from the common area and is placed on the same level as the roof terrace. The other house by Le Corbusier is based on the Citrohan model. It has a double-height living room and anticipates the typical dwelling of Le Corbusier's unites d'habitation.

So it can be seen that both Mies and Corbusier used the Weissenhof exhibition to demonstrate some of their basic ideas. The same was the case with Gropius, who built a technically ingenious house where a light and closely spaced steel frame carried an insulation of cork slabs, and exterior and interior covering of asbestos cement and Celotex sheets respectively. The dry-mounting process introduced here represented an important contribution to the industrialization of building.

From the general urban layout and the planning of dwellings for a modern form of life, to the solution of problems of technology and economy, the Weissen holsiedlung represented a great collective achievement, and immediately became a goal of architectural pilgrimage.

6.3 Vocabulary 词汇表

① asymmetrical adj. 不对称的

② melancholy adj. 忧郁的；令人伤感的

③ enumerate vt. 列举；枚举

④ contour n. 外形，轮廓

⑤ parterres n. 花坛，花圃

⑥ endorse vt. （公开）赞同，支持，认可

⑦ consort n. 配偶（尤指在为君主的配偶）

⑧ fanatical adj. 狂热的

⑨ abhor vt. 对……深恶痛绝；厌恶；憎恶

⑩ articulation n. （正式）思想、感情的表达

⑪ stereometric adj. 立体测量的，立体几何的

⑫ pluralism n. 多元化，多元性

⑬ differentiate vt. 使不同，使有差别

⑭ velocity n. 速度；速率

⑮ longitudinal adj. 纵向的

⑯ promenade n. 步行道

⑰ dynamism n. 动态；活力，干劲

⑱ transparency n. 透明；透明性；透明度

⑲ juxtaposition n. 并置，并列

⑳ Macassar ebony n. 望加锡乌木

㉑ conservatory n. 【英】温室，（玻璃）暖房

㉒ chromium n. 【化】铬

㉓ membrane n. 薄膜，膜状物

㉔ engirdle vt. 以带环绕；围绕

㉕ pilgrimage n. 朝圣；朝觐；参拜圣地

㉖ emphasis n. 重点；强调；加强语气

㉗ untamed adj. 不能抑制的；难控驭的；未驯服的

㉘ picturesque adj. 独特的；生动的；别致的；图画般的

㉙ exotic adj. 异国的；外来的；异国情调的

㉚ asymmetrical adj. ［数］非对称的（等于 asymmetric）；不均匀

的；不匀称的

㉛ irregular n. 不规则物；不合规格的产品 adj. 不规则的；无规律的；非正规的；不合法的

㉜ obsession n. 痴迷；困扰；[内科][心理] 强迫观念

㉝ fragmentation n. 破碎；分裂；[计] 存储残片

㉞ esthetic adj. 审美的（等于 aesthetic）；感觉的 n. 美学；审美家；唯美主义者

㉟ doctrine n. 主义；学说；教义；信条

㊱ vocal adj. 歌唱的；声音的，有声的 adj. 直言不讳的 n. 声乐作品；元音 n. Vocal) 人名；（西）博卡尔

㊲ layout n. 布局；设计；安排；陈列

㊳ Gothic adj. 哥特式的；野蛮的 n. 哥特式

㊴ Revivalist n. 领导宗教复兴运动的人；复旧者

㊵ Augustus Welby Northmore Pugin n. 奥古斯塔斯·普金

㊶ embodiment n. 体现；化身；具体化

㊷ moral adj. 道德的；精神上的；品性端正的 n. 道德；寓意

㊸ enumerated v. 枚举，计算（enumerate 的过去形式）；n. 枚举类型

㊹ ornament n. 装饰；[建][服装] 装饰物；教堂用品 vt. 装饰，修饰

㊺ advocate vt. 提倡，主张，拥护 n. 提倡者；支持者；律师

㊻ secular adj. 世俗的；长期的；现世的；不朽的 n. 修道院外的教士，（对宗教家而言的）俗人

㊼ architect n. 建筑师

㊽ exaggerated adj. 夸张的，言过其实的 v. 夸张，夸大（exaggerate 的过去式）

㊾ rigid adj. 严格的；僵硬的，死板的；坚硬的；精确的

㊿ geometric adj. 几何学的；[数] 几何学图形的

(51) contours n. 轮廓（contour 的复数）；［气象］等高线 v. 画等高线；画轮廓（contour 的三单形式）

(52) fortuitous adj. 偶然的，意外的；幸运的

(53) contrived adj. 人为的；做作的；不自然的

(54) extended adj. 延伸的；扩大的；长期的；广大的 v. 延长；扩充（extend 的过去分词）

(55) grazed adj. 擦破的 v. 擦过；放牧（graze 的过去分词）；抓破

(56) ditch vt. 在……上掘沟；把……开入沟里；丢弃 vi. 开沟；掘沟 n. 沟渠；壕沟

(57) vicinity n. 邻近，附近；近处

(58) concealed adj. 隐蔽的，隐匿的

(59) capabilities n. 能力（capability 的复数）；功能；性能

(60) symmetrically adv. 对称地；平衡地；匀称地

(61) sloping adj. 倾斜的；有坡度的；成斜坡的 v. 溢出（slop 的 ing 形式）

(62) shrubbery n. 灌木；［林］灌木林

(63) overstated v. 夸大；夸张；过分强调（overstate 的过去分词）adj. 夸张的；过高的

(64) transformed v. 彻底改变，使发生巨变，使改观，使改变性质；（transform 的过去式和过去分词）

(65) renowned adj. 著名的；有声望的 v. 使有声誉（renown 的过去分词）

(66) modified adj. 改进的，修改的；改良的 v. 修改；缓和（modify 的过去分词）

(67) merged v. 合并；融合（merge 的过去分词）adj. 合并的；结了婚的

(68) subtle adj. 微妙的；精细的；敏感的；狡猾的；稀薄的

(69) medieval adj. 中世纪的；原始的；仿中世纪的；老式的

第六部分 建筑文本翻译拓展训练（二）汉译英

⑦ pavilions n. 楼阁；大帐篷；观众席；展览馆（pavilion 的复数形式）v. 搭帐篷（pavilion 的三单形式）

⑦ exotics n. 外来植物；超级跑车

⑦ engraving n. 雕刻；雕刻术；雕刻品 n. 雕刻；雕刻术；雕刻品

⑦ Ermenonville 埃尔门维尔

⑦ Frederick Law Olmsted 弗雷德里克·劳·奥姆斯特德

⑦ Calvert Vaux 卡尔弗特·沃克斯

⑦ municipal adj. 市政的，市的；地方自治的

⑦ motifs n. 图案；动机（motif 的复数）

⑦ appropriateness n. 适当；适合

⑦ dimensions n. 规模，大小

⑧ Walpole n. 沃波尔（姓氏）

⑧ silhouette n. 轮廓，剪影 vt. 使……照出影子来；使……仅仅显出轮廓 n. (Silhouette) 人名；（法）西卢埃特

⑧ perimeter n. 周长；周界；[眼科]视野计

⑧ randomly adv. 随便地，任意地；无目的地，胡乱地；未加计划地

⑧ cathedrals 教堂

⑧ erected adj. 直立的；正立的 vt. 建立；竖立（erect 的过去分词）

⑧ abhorred vt. 憎恶；痛恨；（abhor 的过去式和过去分词）

⑧ authentic adj. 真正的，真实的；可信的

⑧ sham n. 假装；骗子；赝品 vt. 假装；冒充 vi. 假装；佯装 adj. 假的；虚假的；假装的

⑧ plaster n. 石膏；灰泥；膏药 vt. 减轻；粘贴；涂以灰泥；敷以膏药；使平服

⑨ vaults n. 拱顶；地窖（vault 的复数形式）；保管库；撑杆跳 v. 给……盖以拱顶；跳跃；跳跃（vault 的第三人称单数形式）

㉛ Pugin 皮金·奥古斯塔斯·韦比·诺思莫尔

㉜ edifices n. 大厦；(edifice 的名词复数)

㉝ Victorian adj. 维多利亚女王时代的；英国维多利亚女王时代的 n. 维多利亚时代；维多利亚女王时代的人

㉞ commissions n. 佣金，手续费；现金奖励情况（commission 的复数）

㉟ monastic adj. 修道院的；僧尼的；庙宇的 n. 僧侣；修道士

㊱ collegiate adj. 大学的；学院的；大学生的

㊲ Clwyd n. 克卢伊德（英国威尔士郡名）

㊳ ornament n. 装饰；[建] [服装] 装饰物；教堂用品 vt. 装饰，修饰

㊴ continuity n. 连续性；一连串；分镜头剧本

⑩ verticality n. 垂直性；[航] 垂直状态

⑪ facade n. 正面；表面；外观

⑫ symmetry n. 对称（性）；整齐，匀称

⑬ panelled adj. 镶框式的；有饰板的 v. 嵌镶（panel 的过去分词）

⑭ awe vt. 使敬畏；使畏怯 n. 敬畏

⑮ impede adj. 必然的，不可避免的

⑯ dwelling n. 住处；寓所 v. 居住（dwell 的现在分词）

⑰ endeavour n. 努力

⑱ skeleton n. 骨架，骨骼；纲要；骨瘦如柴的人 adj. 骨骼的；骨瘦如柴的；概略的

⑲ stereometric 立体测量学

⑳ volumes n. 卷；大量；音量；体积；(volume 的名词复数)

㉑ motifs n. 动机；主题；意念；装饰图案；（motif 的复数）

㉒ creed n. 宗教信条

㉓ devalued vt. 贬低；（devalue 的过去式）vi. 贬值

⑭ elaborate v. 详细地说明

⑮ chaotic adj. 混沌的；混乱的，无秩序的

⑯ pedestrian adj. 徒步的；缺乏想像力的 n. 行人；步行者

⑰ velocities n. 速度（velocity 的复数）

⑱ stilts n. 高跷；支柱；（stilt 的复数）

⑲ spatial adj. 空间的；存在于空间的；受空间条件限制的

⑳ storeys n. 楼层；叠架的一层；（storey 的复数）

㉑ culminate vi. 达到顶点；达到高潮；（culminate 的过去式）vt. 使结束

㉒ corridor n. 走廊；过道

㉓ incomparable adj. 无双的；不能比较的；截然不同的

㉔ villa n. 乡间别墅；（Villa）人名：比利亚

㉕ rigour n. 严密；精确

㉖ humidity n. ［气象］湿度；湿气

㉗ utilized v. 利用；（utilize 的过去式和过去分词）

㉘ longitudinal adj. 纵向的；经度的

㉙ axis n. 轴线；坐标轴；中心部分；枢椎；联盟；斑鹿

㉚ splendid adj. 壮丽的；宏伟的

㉛ ramp n. 斜面；弯子；斜坡波形；诈骗 v. 为……提供斜面；前肢腾空而起；电压随时间增减；购进股票

㉜ transparency n. 透明，透明度；幻灯片；有图案的玻璃

㉝ promenade n. 散步场所；同 PROM v. 散步兜风

㉞ sphere n. 圆形实心体；球面；范围；社会阶层 v. 将……封入球体中

㉟ cylinder n. 圆柱体

㊱ protrusion n. 突出；突出物

㊲ accommodates vt. 使适应；容纳；调解 vi. 适应；调解

㊳ Ludwig Mies 路德维希·密斯

⑬⑨ der Rohe 德·罗厄

⑭⓪ Le Corbusie 勒柯布西耶

⑭① prototype v. 制作（产品的）原型；n. 原型；典型

⑭② skyscrapers n. 摩天大楼；（skyscraper 的名词复数）

⑭③ reminiscent adj. 怀旧的，回忆往事的；耽于回想的 n. 回忆录作者；回忆者

⑭④ Villa Savoye 萨沃耶别墅

⑭⑤ terrace n. 露台；排屋 v. 使成阶地

⑭⑥ rectangular adj. 长方形的；成直角的

⑭⑦ podium n. 讲台

⑭⑧ Onyx n. 缟玛瑙

⑭⑨ ebony n. 乌木；乌木树

⑮⓪ chromium n. 铬

⑮① Weimar n. 魏玛（德国城市）

⑮② manifest vt. 证明，表明；显示 vi. 显示，出现 n. 载货单，货单；旅客名单 adj. 显然的，明显的；明白的

⑮③ dynamic adj. 动态的；动力的；动力学的；有活力的 n. 动态；动力

⑮④ sham n. 假装；骗子；赝品 vt. 假装；冒充 vi. 假装；佯装 adj. 假的；虚假的；假装的

⑮⑤ axial adj. 与轴有关的

⑮⑥ balconies n. 阳台；露台；（戏院）包厢（balcony 的复数形式）

⑮⑦ light-modulator 光调制器

⑮⑧ Bauhaus 包豪斯建筑学院

⑮⑨ Gropius 格罗皮乌斯

⑯⓪ insulation n. 绝缘；隔离，孤立

⑯① slab n. 平板；厚片 v. 把……切成厚片；用平板盖上

⑯₂ asbestos n. 石棉 adj. 石棉的

⑯₃ cement vt. 巩固，加强；用水泥涂；接合 vi. 粘牢 n. 水泥；接合剂

6.4 Syntactic Structure 重点句型分析

① 英国哥特复兴的领军人物奥古斯塔斯·普金认为哥特式是过去时代的道德和宗教价值的体现，而他自身所处的时代却匮乏的。

In England the leading Gothic Revivalist was Augustus Welby Northmore Pugin (1812 – 1852), who regarded the Gothic as an embodiment of proper moral and religious values from the past that he felt were all but absent in his own time.

解析：who regarded the Gothic as an embodiment of proper moral and religious values from the past that he felt…为非限制定语从句修饰 Augustus Welby Northmore Pugin。定语从句中，regard…as 译为"把……视作"，as 的宾语为 an embodiment of proper moral and religious values from the past，that he felt were all but absent in his own time 为定语从句修饰 proper moral and religious values from the past。

② 同样地，连透过窗子欣赏到的景致都做了同样巧妙的设计，由近处层叠的树木，推及远处吃草的牛儿，凡目光所及均可如画。

Likewise, equally carefully planned views from the windows of buildings extended beyond nearby clumps of trees into the more distant landscape, where cows grazed as in a landscape painting.

解析：全句的主语为 equally carefully planned views，定语从句 where cows grazed as in a landscape painting 修饰 the more distant landscape。

③ 在布伦海姆宫和查兹沃斯这样大规模的园林中，布朗营造出

一番野趣,你见不到对称布局的花坛和笔直的步道,取而代之的是开阔的草坪,连绵的小丘上曲折的山路,点缀着自然种植的树木的溪流和池塘。

At such great houses as Blenheim Palace and Chatsworth, Brown replaced the parterres of symmetrically arranged flower beds and straight walks with sweeping lawns, sloping hills with curving paths, and rivers and ponds punctuated by informally planted groups of trees and shrubbery, to achieve the effect of a wilderness.

解析:replace...with 指"用……取代,代替",punctuated by informally planted groups of trees and shrubbery 为过去分词短语作后置定语。

④ 普金捍卫哥特式,源于哥特式不仅是"一种风格更是一种原则"。在普金看来,哥特式之所以是"真实的",是因为它诚实地使用建筑材料,结构暴露出来,功能由此而得以展示。

Pugin's defence of his chosen style, on the grounds that it was "not a style but a principle", involved the claim that Gothic was "true" because it was the result of an honest use of materials in which structure was exposed and function thereby demonstrated.

解析:on the grounds that it was "not a style but a principle" 为插入语,其前为句子的主语,其后为句子的谓语。in which structure was exposed and function thereby demonstrated 为定语从句。

⑤ 他的著作对整个维多利亚时期哥特式建筑的设计和建造产生了深远的影响,比他同时代所有人的贡献都大,甚至超越他本人的建筑作品,这些作品常常饱受资金缺乏之苦。

His books exerted a lasting and wide-ranging influence on the Gothic architecture designed and built throughout the whole Victorian period, greater than almost any other architect's contribution at this time and certainly greater than Pugin's own architectural work, commissions which often suffered from lack of adequate financing.

解析：designed and built throughout the whole Victorian period 为过去分词短语作后置定语，两个 greater than 引导的成分主要主语补足语，补充说明句子的主语 His books。commissions which often suffered from lack of adequate financing 为同位语，说明 Pugin's own architectural work 的具体内容。

⑥ 功能主义建筑的基本形式和严格原则可以解释成这样一种主张：反对建筑师们所谓历史主义的廉价母题和学究式句法，但首先所有的功能主义者都有一个基于对人类和建筑的强烈信仰之上的积极目标。

The elementary forms and strict principles of Functionalist architecture may be interpreted as a protest against what the architecture considered to be the devalued motifs and academic compositions of historicism, and it possessed the significance and power to become an international movement.

解析：全句主语 The elementary forms and strict principles of Functionalist architecture，介词 against 的宾语为 what 从句，what 在从句中做 considered 的宾语。

⑦ 他由此认识到框架结构和开放空间之间的天然联系，从而使这些仅仅被少数先锋建筑师直觉感受到的东西变成普遍可能。

He thus recognized the natural relationship between skeleton construction and open space and made generally possible something that so far had only been intuited by a few pioneers.

解析：made 的宾语为 something，作为宾语补足语的 generally possible 之所以提前是因为相较于带定语从句的宾语过短，为了句子的平衡而做的调整。

⑧ 它不仅是个实践性理念，它的出发点是不同交通工具意味着不同的运动节奏，因而需要不同空间的认识。

The idea was not merely a practical one but took as its point of departure the recognition that different velocities imply different rhythms, and hence require different kinds of space.

解析：took...as 指"把……看作"，took 的宾语为由 that 从句修饰的 the recognition。

⑨ 通向主要楼层的坡道本身就是一个用空间来适应不同功能的巧妙试验，坡道继续向上，可以到达屋顶的"日光浴室"，在那里，曲线的墙体综合了随意，私密以及周围景观的相互作用。

The ramp leads up to the main floor which is a masterly exercise in spatial adaptation to various functions, and from there on to the "solarium" on the roof, where curved walls represent a synthesis of informality, privacy and interaction with the surrounding landscape.

解析：which，where 分别引导定语从句及非定语从句。

⑩ 萨伏伊别墅空间的丰富和动感完全被收容在一个近似方形的体量中，这实现了勒·柯布西耶的两个基本意图：对空间自由度的要求和对基本形式的要求。

The spatial richness and dynamism of the Villa Savoye is contained within an approximately square volume, which allows Le Corbusier to satisfy both of his basic intentions: the desire for spatial freedom and the desire for elementary form.

解析：句子主语 The spatial richness and dynamism of the Villa Savoye，which 引导非限制定语从句，the desire for spatial freedom and the desire for elementary form 为 both of his basic intentions 的同位语成分。

⑪ 其中一半被覆盖于与方盒子相连的屋顶之下，这屋顶的一部分由钢柱支撑着，体现了主要层结构系统的连续性。

It is half covered by a roof connecting the boxes, and partially resting on steel columns, which represent a continuation of the structural system of the main floor.

解析：It 指代 the space，connecting 和 resting 为现在分词短语修

饰 a roof，which 为非限制定语从句的主语。

⑫ 为了理解密斯的空间表达性处理，我们必须研究这些不同类型的墙如何交接在一起，因为所有的交集点和细部都有着组织空间的功能，并且体现着潜藏其中的开放空间的普遍概念。

To understand Mies's spatial articulation, it is important to study how the different types of walls are joined together, for all joints and details are a function of the spatial composition and express the general underlying concept of open space.

解析：To understand Mies's spatial articulation 做目的状语，it 为形式主语，句子真正的主语为 to study how the different types of walls are joined together，for 引导愿意状语从句。

⑬ 其中每一部分都占据单独的一翼，整个平面表现了对功能分区的典型愿望，同时实现了建筑与环境的积极联系。包豪斯的开放和动态的形式由三个主要体块间的连接元素进一步加强。

Each of these was housed in its own wing, a plan that demonstrated a typical wish for functional differentiation and at the same time created an active relationship between the building and its surroundings.

解析：that 引导定语从句修饰 a plan，定语从句中有两个谓语动词 demonstrated 和 created。

⑭ 尽管这里有着使用共同的形式语言的明确要求，但建筑师的不同个性还是十分明显：从密斯精确的优雅，到奥德和斯塔姆克制的简约，以及夏隆的动态体量。

Although there is a noticeable wish for a common formal language, the architectural character varies considerably from the calculated elegance of Mies to the restrained simplicity of Oud and Stam, and the dynamic volumes of Scharoun.

解析：句子的谓语用 varies from…to…来表示建筑特征的多样性。

⑮ 他设计了一个有着技术独创性的住宅，它由一个轻巧的布置

紧凑的钢框架支撑起厚软木的保温墙，室外和室内分别铺设了石棉板和隔音板。

The same was the case with Gropius, who built a technically ingenious house where a light and closely spaced steel frame carried an insulation of cork slabs, and exterior and interior covering of asbestos cement and Celotex sheets respectively.

解析：where 引导定语从句，从句中 a light and closely spaced steel frame 作主语。

简明建筑英语
翻译教程
Chapter 7

第七部分　建筑学常用专业术语汉英对照

城市史相关术语

城市一般术语

城邦 polis

城堡 castle

城壕 city moat, city trench

城［墙］city wall

城市 city

城市改建 urban redevelopment

城市更新 urban renewal

城市历史保护 urban conservation

城市史 history of cities

大城市 large city

大学城 campus city

都城 capital city

风景旅游城市 sceic－tourist city

港口城市 port city

工业城市 industrial city

宫城 imperial city

关厢 outskirt

胡同 hutong, alley

皇城 imperial palace

郊区 suburb

街 street

居民点 settlement

科学城 science city

矿业城市 mining city

邻里 neighbourhood

历史地段 historical site

历史建筑 historical building

商业城市 commercial city

市［场］market

市区 city proper

外城 outer city

巷 lane

小城市 small city

休养城市 resort city

镇 township，town

中等城市 medium – sized city

外国古代建筑及构造术语

阿拉伯花饰 arabeque

阿拉伯式柱头 Arabian capit

半穹顶 semi dome

半球形穹顶 semispherical dome

半圆券 semicircular arch

半四形壁龛 apse

壁上拱廊 blind arcade

采光塔 lantern

槽沟 flute

侧廊 aisle

葱形穹顶 ogee dome

葱形尖拱 ogee arch

倒槽式拱 trough vault

滴水兽 gargoyle

第七部分 建筑学常用专业术语汉英对照

顶端饰 finial

多立克柱式 Doric order

额枋 architrave

帆拱 pendentive

方尖碑 obelisk

方穹顶 square dome

扶壁 buttress

府邸 mansion, palazzo

拱廊 atcade

拱式 vault type

鼓座 drum

广场 forum

弧形券 segmental arch

花色窗棂 tracery

会堂 basilica

尖券 pointed arch

尖项 pinnacle

讲坛 mimbar, pulpit

交角券 squinch

金字塔 pyramid

角斗场 colosseum

凯旋门 triumphal arch

科林斯柱式 Corinthian order

肋骨拱 rib bed vault

棱拱 groin vault

莲花式柱 lotus coiumn

列柱围廊式 peristyle

马蹄券 horseshoe arch

玫瑰窗 rose window

门廊 potch, portico

牌门楼 pylon

平嵌线 reglet

穹式 dome type

券式 arch type

山花 pediment

扇形拱 fan vault

神坛 altar

圣龛 mihrab, prayer niche

输水道 aqueduct

四叶饰 quatrefoil

塔楼 turret

塔庙 ziggurat

台基 stylobate

筒拱 barrel vault

托斯卡柱式 Tuscan order

万神庙 pantheon

网状拱 reticulated vault

涡卷装饰 cartouche

五叶饰 cinquefoil

小侧窗 buleye window

小穹顶 cupola

星形拱 stellar vault

袖廊 transept

宣礼塔 minaret

檐壁 frieze

檐部 entablature

檐口 cornice

伊奥尼亚柱式 Ionic order

圆底线脚 torus

纸草花式柱 papyrus column

中厅 central nave

钟乳拱 stalactite vault

主穹顶 main dome

柱基 plinth

柱廊 colonnade

柱身 column shaft

柱头 capital

棕榈叶式柱 palm column

中国古代建筑及构造术语

昂 ang, lever

八椽栿 9 - purlin beam（宋代术语）

拔檐 corbel, hanging over

板瓦 flat tile

抱鼓石 drum - shaped bearing stone

抱框 jamb on door or window

抱头梁 baotou beam

碑碣 stone tablet

博古架 antique shelf

博脊 horizontal ridge for gable and hip roof

搏缝板 gable eave board

搏缝版 gable eave board（宋代术语）

补间铺作 bracket sets between columns（宋代术语）

步架 horizontal spacing between purlins

材 cai

彩画 colored pattern, colored drawing

槽升子 center block

侧脚 cejiao（宋代术语）

叉手 chashou, inverted V-shaped brace（宋代术语）

撑头 small tie-beam

鸱尾 chiwei（宋及宋前术语）

重檐 double eave roof

出 extension of bracket

出跳 extension of bracket（宋代术语）

穿插枋 penetrating tie

穿斗式构架 column and tie construction

垂带 drooping belt stone

垂脊 diagonal ridge for hip roof, vertical ridge for gable roof

垂台钩栏 double frieze balustrade

攒 set of bracket

攒尖 pyramidal roof

大斗 cap block

大木 wooden structure

大木作 carpentry work

大式 wooden frame with dougong

单步梁 one-step cross beam

单材拱 outer-side bracket arm

单钩栏 single frieze balustrade（宋代术语）

滴水 drip tile

地伏 plinth stone

点金 gold pointing

殿 hall

叠涩 corbel, hanging over（宋代术语）

顶椽 top rafter

斗 dou, bracket set

斗板石 intermediate pier

斗拱 dougong, bracket set

斗口 doukou, mortise of cap block

额枋 architrave（used with dougong）大式

枋 tiebeam

枋 small tie-bean（宋代术语）

枋心 central portion of painted beam

房 house

舫 boat house

飞檐椽 flying rafter

飞子 flying rafter（宋代术语）

分 fen

趺 fu, stone turtle（清代术语）

栿 beam（宋代术语）

副子 drooping belt stone（宋代术语）

阁 pavilion

格栅 partition door

隔断墙 vertical parition wall

拱 gong, bracketarm

勾头 eave tile

钩栏 balustrade（宋代术语）

箍头 end portion of painted beam

瓜拱 oval armn

瓜柱 short column

瓜子拱 oval arm（宋代术语）

圭角 guijiao

海墁天花 flat ceiling

领版 tile edging（宋代术语）

和玺彩画 dragons pattern

花架椽 intermediate rafter

华版 frieze panel

华表 huabiao, ornamental pillar

华拱 flower arm（宋代术语）

脊枋 ridge tiebeam

脊瓜柱 king post

脊檩 ridged purlin 小式

脊（木行）ridged purlin 大式

脊抟 ridged purlin（宋代术语）

脊瓦 ridge tile

槛 stud

槛窗 sill wall window

槛墙 sill wall

交互斗 connection block（宋代术语）

角背 bracket

角科 bracket set on corner

角梁 hip rafer

角柱 corner column（唐、宋术语）

角柱石 corner pier

阶条石 rectangular stone slab

金檩 intermediate purlin 小式

金（木行）intermediate purlin 大式

金柱 hypostyle column

井干式构架 log cabin construction

井口天花 compartment ceiling

九架梁 9 – purlin beam

举架 raising the purlin

举析 raising the purlin（宋代术语）

卷棚 round ridge roof

卷杀 entasis

龛 niche

框槛 door frame

栏版 frieze panel

阑额 architrave（宋代术语）

廊 colonnade

廊墙 partition wall

老檐（木行）purlin on hypostyle 大、小式

老檐枋 eave tiebeam 大式，指檐口构造

雷公柱 suspended column

沥粉 embossed painting

连檐 eave edging

梁 beam

檩 purlin（used without dougong）小式

棂子 grill

令拱 regular arm（宋代术语）

琉璃瓦 glazed roof tile

六椽栿 7 – purlin beam（宋代术语）

六架梁 6 – purlin beam

楼 storied building（"某某楼"一般译为 tower，如"岳阳楼"译为 Yueyang Tower）

栌斗 cap block（宋代术语）

慢拱 long arm（宋代术语）

帽儿梁 lattice framing

门跋 door knocker

门钉 decorative nails on door leaf

门簪 decorative cylinder

门枕 door bearing

门枕石 bearing stone

民居 folk house

抹头 window stool

木装修 joiner's work

脑椽 upper ralter

内槽柱 hypostyle column（唐代术语）

内檐装修 interior finish work

内柱 hypostyle column（宋代术语）

泥道拱 axial bracket arm（宋代术语）

牌楼 pailou，decorated gateway

平梁 3 – purlin beam（宋代术语）

平身科 bracket sets between columns

七架梁 7 – purlin beam

齐心斗 center block（宋代术语）

栔 qi

羌差 ramp

戗脊 diagonal ridge for gable and hip roof

翘 flower arm，petal

青瓦 black tile

雀替 sparrow brace

阙 que，watchtower

裙板 panel

乳栿 rufu，beam tie（宋代术语）

三才升 small block

三架梁 3-purlin beam

山柱 center column

扇面墙 horizontal partition wall

上枋 upper fillet and fascia

上金枋 upper purlin tiebeam

上枭 upper cyma, recta

上中平抟 intermediate purlin（宋代术语）

升 block with cross mortise

十八斗 connection block

石人 stone human statue

石像生 stone animal

蜀柱 king post（宋代术语）

束腰 suyao

耍头 nose

双步梁 two-step cross beam

水榭 waterside pavilion

四阿 hip roof

四椽栿 5-purlin beam（宋代术语）

四合院 courtyard house

四架梁 4-purlin beam（宋代术语）

苏式彩画 Suzhou style pattern

梭柱 shuttle-shaped column（宋代术语）

塔 pagoda

榻板 window sill

踏 step（宋代术语）

踏道 steps（宋代术语）

踏垛 step

台 platform

台阶 steps

抬梁式构架 post and lintel construction

坛 altar

天花 ceiling

天花彩画 ceiling pattern

挑尖梁 main aisle exposed tiebeam

挑檐石 cantilever stone on eave

跳 tiao（宋代术语）

贴金 gold foil painting

厅 hall（又称堂、室）

亭 pavililon

筒瓦 round tile

抟 purlin（宋代术语）

退晕 color change

驼峰 tuofeng, camel-hump shaped suport（宋代术语）

柁墩 wooden pier

瓦当 tile end, eave tile with pattern

瓦口 tile edging

瓦石作 masonry

外檐装修 exterior finish work

万拱 long arm

望柱 baluster

望柱头 baluster capital

五花山墙 stepped gable wall

五架梁 5-purlin beam

庑殿 hip roof

榍头 gable wall head

贔 bixi, stone turtle（宋及宋前术语）

下枋 lower fillet and fascia

下金枋 lower purlin tiebeam

下枭 lower cyma, reversa

下平抟 eave purlin（宋代术语）

仙人 celestial being

厢拱 regular arm

象眼 triangular space

小木作 joinery work

小式 wooden frame without dougong

歇山 gable and hip roof

榭 pavilion on terrace

须弥座 xumizuo

轩 windowed veranda

悬山 overhanging gabie roof

旋子彩画 tangent circle pattern

寻杖 handrail（宋代术语）

压阑石 rectangular stone slab（宋代术语）

檐椽 eave rafter

檐垫板 cushion board 小式

檐枋 architrave（used with dougong）小式

檐枋 eave tiebeam 小式，指檐口构造

檐檩 eave purlin 小式

檐墙 eave wall

檐柱 eave column

硬山 flush gable roof

由额垫板 cushion board 大式

由戗 inverted V–shaped brace

御路 yulu, imperial path

圆攒尖 round pavilion roof

月梁 crescent beam

藻井 caisson ceiling

藻头 intermediate portion of painted beam

集 stockaded village

帐杆 suspended column（宋代术语）

罩 shelf

正脊 main ridge

正吻 zhengwen

正心（木行）eave purlin

大式正心拱 axial bracket arm

支条 lattice framing

支摘窗 removable window

踬 zhi

柱础 column base

柱顶石 capital stone

柱头科 bracket set on columns（宋代术语）

柱头铺作 bracket set on columns

砖雕 brick carving

转角铺作 bracket set on corner（宋代术语）

走兽 animal

中国古典园林史名词

阿房宫 E-Pang Palace（秦代术语）

暗轴 hidden axis, invisible axis（又称"隐轴"）

巴洛克式园林 Baroque style garden

背景 background

苍古 antiquity

承德避暑山庄 Chengde Imperial Summer Resort

帝王宫苑 imperial palace garden

动观 in-motion viewing

对景 oppositive scenery, view in opposire place

法兰西式园林 French style garden

凡尔赛宫苑 Versailles Palace Park

枫丹白露宫园 Fontainebleau Palace Garden

封闭空间 encloseure space

副景 secondary feature

副轴 auxiliary axis

艮岳 Gen Yue Imperial Garden（宋代术语）

华清宫 Hua-Qing Palace（唐代术语）

皇家园林 royal garden

夹景 vista line, vista（又称"风景线"）

江南园林 garden on the Yangtze Delta

借景 borrowed scenery, view borrowing

近景 nearby view

景 view, scenery, feature

景点 feature spot, view spot

景序 order of sceneries

静观 in-position viewing

开敞空间 wide open space, wide space

空灵 spaciousness, airiness

框景 enframed scenery

廊柱园 peristyle garden, patio

勒诺特尔式园林 Le Notre's style garden

灵台 Ling Tai Platform Garden（周代术语）

灵囿 Ling You Hunting Garden（周代术语）

灵沼 Ling Zhao Water Garden（周代术语）

漏景 leaking through scenery

绿廊 xystus

洛可可式园林 Rococo style garden

洛阳宫 Luoyang Palace（魏代术语）

迷阵 maze, labyrinth

配景 objective view

前景 front view

上林苑 Shang-Lin Yuan（汉代术语）

私家园林 private garden

苏州园林 Suzhou traditional garden

缩景 miniature scenery, abbreviated scenery

未央宫 Wei Yang Palace（汉代术语）

尾景 terminal feature

文艺复兴庄园 Renaissance style garden

西班牙式园林 Spanish style garden

西方古典园林 western classical garden

悬园 Hanging Garden（又称"悬空园""架高园"）

颐和园 Yi-He Yuan Imperial Garden

意大利式园林 Italian style garden

意境 artistic conception, poetic imagery

英国皇家植物园 Royal Botanical Garden, Kew garden（又称"邱园"）

英国式园林 English style garden

园林空间 garden space

圆明园 Yuan-Ming Yuan Imperial Garden

远景 disrant view

障景 obstructive scenery, blocking view

中国传统园林 traditional Chinese garden

中国古代园林 ancient Chinese garden

中国古典园林 classical Chinese garden

中国山水园 Chinese mountain and water garden

中英混合式园林 Anglo-Chinese style garden

轴线 axis, axial line

主景 main feature

主轴 main axis

庄园 manor, villa garden

工业与民用建筑术语

安全出口 safe exit

安全抓杆 grab bar

包厢 box

包厢 box (in the auditorium)

避难走道 fire-protection evacuation walk

边幕 wings

殡仪馆 funeral parlor

殡仪区 division for funeral service

冰道 swimming lane

侧台 bay area

查阅档案用房 search room

敞开式汽车库 open garage

敞开式汽车库存 open garage

池座 stalls

出发台 starting block

大幕 proscenium curtain

单体厕所 monocase public toilets

单元式高层住宅 tall building of apartment

档案馆 archives

档案库 storehouse for archi

岛式舞台 arena stage

悼念厅 mourning hall

灯光吊笼 lighting (cable) basket

灯光渡桥 lighting bridge

灯控室 lighting control room

低位小便器 tow – level urinal

地道工程 undermined works without low exit

地下街 underground street

地下汽车库 under ground garage

典藏室 book – keeping department

吊杆 batten

独立式公共厕所 independence pubic toilets

多层厂房（仓库） multi – storied industrial building

耳光室 fore stage side lighting

发车位 seat of delivery passenger vehicle

防爆地漏 blastproof floor drain

防毒通道 air – lock

防护单元 protective unit

防护密闭门 airtight blast door

防化通信值班室 CBR protection and communication duty room

防空地下室 air defence basement

防空专业队工程 works of service team for civil air defence

防烟楼梯间 smoke prevention staircase

飞机库 aircraft hangar

飞机停放和维修区 aircraft storage and servicing area

封闭外廊 closed corridor

辅助书库 auxiliary stacks

附属式公共厕所 dependence public toilets

复式汽车库 compound garage

高层厂房（仓库）high–rise industrial building

高层汽车库 high–rise garage

高等学校图书馆 college library

高级旅馆 high–grade hotel

高架仓库 high rack storage

高架跨线候车室 elevated overcrossing waiting room

公共厕所 public toilets, lavatory, restroom

公共活动室 activity room

公共建筑 public building

公共图书馆 public library

公交站点 bus stop

固定式公共厕所 fix up public toilets

固定坐席 fixing seating

观众厅 auditorium

观众席 seats for the spectator

国家级档案管 national archives

后郎台 back stage

护理院 nursing home

缓坡段 transition slope

活动式公共厕所 mobile public toilets

活动坐席 retractable seating

机械式立体汽车库 mechanical and stereoscopic garage

机械式汽车库 mechanical garage

积层书架 stack–system shelf

基本书库 basic stack rooms

假台口（或活动台口）movable pretendnd stage door

竞赛区 arena

镜框式舞台 proscenium stage

居住建筑 residential building

剧场 theater

开敞式舞台 open stage

开架书库 open stacks

开架阅览室 open stack reading room

看台 stands

科学研究图书馆 research institution library

坑道工程 undermined works with low exit

口部建筑 gateway building

老年人公寓 apartment for the aged

老年人居住建筑 residential building for the aged

老年人住宅 house for the aged

老年人住宅 house for the aged

乐池 orchestra pit

两层式机械汽车库 two storey mechanical garage

流动灯光 movable lighting

楼座 balcony

旅客车站专用场地 special area for passenger station

轮椅坡道 ramp for wheelchair

轮椅通道 passage for wheelchair

轮椅席位 seat for wheelchair

盲道 sidewallk for the blind

密闭通道 airtight passage

密集书架 compact bookshelf

密集书库 compact stacks

面光桥 fore stage lighting gallery

民用建筑 civil building

明步楼梯 stairs without barricade

母片库 storehouse for master

跑道 track

配套工程 indemnificatory works

坡道式汽车库 ramp garage

普通阅览室 general reading room

汽车厕所 bus public toilets

汽车客运站 bus passenger terminal

汽车库 garage

汽车最小转弯半径 minimumturn radius of car

前檐幕 fore-proscenium curtain

热身场地 warming up area

人防围护结构 surrounding structure for civil air entrance

人防有效面积 effective floor area for civil air defence

人民防空地下室 civil air defence basement

人民防空工程 civil air defence works

人员遮蔽工程 personnel shelter

纱幕 veil curtain

商业服务网点 commercial service facilities

商业服务网点 commercial serving

商住楼 business-living building

伸出式舞台 thrust stage

升降池底 adjustable floor

升降乐池 orchestra lift

升降平台 lift platform

升降台 elevating stage

声控室 sound control room

使用面积 usable area

书架层 stack layer

疏散出口 evacuation exit

疏散走道 evacuation walk

竖直循环式机械汽车库 vertical circular garage

水下音响 underwater sound

水下照明 underwater lighting

塔式高层住宅 apartment of tower building

台唇 apron stage

台口 proscenium opening

台口墙轴线 axis of proscenium wall

台口线 curtain line

台口柱光架 lighting tower

台塔 fly tower

特藏书库 special stacks

特种阅览室 special reading room

提示盲道 warming blind sidewalk

体育场 stadium

体育馆 sports hall

体育建筑 sports building

体育设施 sports facilities

天幕 cyclorama

天桥侧光 fly gallery lighting

跳水池 diving pool

铁路客运站 railway passenger station

停车场 parking area

停车位 parking space

通廊式高层住宅 gallery tall building of apartment

托老所 nursery for the aged

拖动厕所 drag movable public toilets

弯道超高 ramp turn supperelcvation

文献资料 document literature

无障碍厕所 barrier – free lavatory

无障碍厕位 barrier – free toilet cubical

无障碍电梯 barrier – free lift

无障碍淋浴间 barrier – free shower room

无障碍盆浴间 barrier – free bath room

无障碍入口 barrier – free entrance

无障碍设施 accessbility facilities

无障碍住房 barrier – free residence

无障碍专用厕所 toilets for disable people

舞台 stage

舞台监督指挥系统 stage manager control system

舞台监视系统 stage monitoring system

信息处理用房 information processing room

兴奋剂检测室 doping control room

行包装卸廊 package transportation corridor

行进盲道 go – ahead blind sidewalk

修车库 motor repair shop

训练池 training pool

训练房 practice room

檐幕 transverse curtain

养老院 rest home

业务区 division tor business

医疗救护工程 works of medical treatment and rescure

游泳池 swimming pool

游泳设施 natatorial facilities

缘石坡道 curb ramp

跃层住宅 duplex apartment

站场客运建筑 buildings for passenger taffic in station yard

站房平台 platform for station building

站前广场 blaz before the station

珍藏库 storehouse for precious archives

珍贵档案 precious archives

重要的办公楼 important office building

重要公共建筑 important public building

主台 main stage

住宅安全性能 residential building safety

住宅单元 residential building unit

住宅环境性能 residential building environment

住宅建筑 residential building

住宅经济性能 residential building economy

住宅耐久性能 residential building durability

专门档案馆 special archives

专门图书馆 special library

转台 revolving stage

综合楼 multiple-use building

综合性档案馆心 comprehensive archives

走无障碍通路 barrier free passage

足球场 football pitch

组装厕所 movable combination public toilets

绿色建筑与智能建筑术语

办公自动化系统 office automation system（OAS）

采暖度日数 heating degree day based on 18℃（HDD18）

采暖年耗电量 annual heating electricity consumption

参照建筑 reference building

非传统水源 nontraditional water source

分散供热水系统 individual hot water supply system

集热器倾角 tilt angle of collector

集热器总面积 gross collector area

集中—分散供热水系统 colectice individual hot water supply system

集中供热水系统 collective hot water supply system

建筑设备自动化系统 building automation system（BAS）

建筑物耗冷量指标 index of cool loss of building

建筑物耗热量指标 index of heat loss of building

可见光透射比 visible transmittance

可再利用材料 reusable material

可再生能源 renewable energy

可再循环材料 recyclable material

空调采暖年耗电量 annual cooling and heating electricity comsumption（EC）

空调采暖年耗电指数 annual cooling and heating electricity consumption factor（ECF）

空调、采暖设备能效比 energy efficiency ratio（EER）

空调度日数 cooling degree day based on 26℃（CDD26）

空调年耗电量 annual cooling electricity consumption

绿色建筑 green building

平板型集热器 flat platec collector

平均窗墙面积比 mean ratio of window area to wall area

强制循环系统 forced ciculation system

热惰性指标（D）index of thermal inertia

日照标准 insolation standards

太阳辐照量 solar irradiation

太阳能保证率 solar fraction

太阳能集热器 solar collector

太阳能间接系统 solar indirect system

太阳能热水系统 solar water heating system

太阳能直接系统 solar direct system

通信网络系统 communication network system（CNS）

透明幕墙 transparent curtain wall

外窗的综合遮阳系数 overall shading coefficient of window

系统集成 systems integration（SD）

真空管集热器 evacuated tubec collector

直流式系统 series-connected system

智能化系统 intelligence system

智能建筑 intelligent building（IB）

贮热水箱 heat storage tank

自然循环系统 natural circulation system

综合布线系统 generic cabling system（GCS）

综合部分负荷性能系数 integrated part load value（IPLV）

建筑设计术语

安装基准面 erection datum plane

半地下室 semi-basement

避难层 refuge storey

变形缝 deformation joint

标准层 typical floor

储藏空间 store space

地下室 basement

垫层 under layer

吊顶 suspended ceiling

吊柜 wall hung cupboard

调整面 coordination face

防潮层 moisture-proof course

辅助基准面 sub-datum plane

隔离层 isolating course

公共绿地 public green space

管道井 pipe shaft

横向缩缝 crosswise stretching crack

回廊 cloister

基准点 datum point

基准面 datum plane

准线 datum line

架空层 open floor

架空层 empty space

架空走廊 bridge way

建筑控制线 building line

建筑密度 building density; building coverage ratio

建筑模数 construction module

建筑幕墙 building curtain wall

结合层 combined course

居住空间 habitable space

勒脚 plinth

绿地率 greening rate

落地橱窗 french window

门斗 foyer

面层 surface course

模数层高 modular storey hight

模数空间网格 modular space grid

模数楼盖高度 modular floor hight

模数室内高度 modular room hight

模数网格 modular grid

模数协调 modular coordination

平台 terrace

平屋层高 storey height

坡屋面 sloping roof

容积率 plot ratio, floor area ratio

入口平台 entrance platform

设备层 mechanical floor

设计使用年限 design working life

伸缝 stretching crack

使用面积 usable area

室内净高 interior net storey height

缩缝 shrinkage crack

套 dwelling space

套型 dwelling size

填充层 filler course

挑廊 overhanging corridor

眺望间 view room

通风道 air relief shaft

围护结构 envelop enclosure

围护性幕墙 enclosing curtain wall

烟道 smoke uptake; smoke flue

檐廊 eaves gallery

永久性顶盖 permanent cap

永久性顶盖 permanent cap

用地红线 boundary line of land; property line

雨篷 canopy

找平层 troweling course

中间层 middle – floor

中间平台 landing

主通道 Main Passageway

装饰性幕墙 decorative faced curtain wall

自然层 floor

自然层数 natural storeys

纵向缩缝 lengthwise shrinkage crack

走廊 corridor gallery

建筑给水排水设计术语

暗设 concealed installation, embedded installation

闭式热水供应系统 closed system for hot water supply

并联供水 parallel water supply

重现期 recurrence interval

串联供水 series water supply

单管热水供应系统 one – pipeline hot water system

地面集水时间 inlet time

第一循环系统 heat carrier circufation system

二级处理 secondary treatment

隔油池 grease interceptor

给水系统 water supply system

工业企业用水 demand for industrial use

供水量 supplying water

官网漏损水量 leakage

化粪池 septic tank

回水管 return pipe

汇水面积 catchment area

集中热水供应系统 central hot water supply system

间接排水 indirect drain

建筑物中水 reclaimed water system for building

建筑中水 reclaimed water system for buildings

降温池 cooling tank

降雨历时 duration of rainfall

降雨强度 rainfall intensity

浇洒道路用水 street flushing demand, road watering

接户管 building unite pipe

径流系数 runoff coefficient

居民生活用水 demand in households

局部热水供应系统 local hot water supply system

开式热水供应系统 open system for hot water supply

绿地用水 green belt sprinkling, green plotsprinkling

埋设深度（覆土深度） buried depth

明设 exposed installation

取水构筑物 intake structure

日变化系数 daily variation coefficient

入户管（进户管） inlet pipe

设计小时耗热量 design heat consumption of maximum hour

时变化系数 hourly variation coefficient

竖向分区 vertical division block

同程热水供应系统 reversed return hot water system

未预见用水量 unforeseen demand

小区中水 reclaimed water system for residential district

悬吊管 hanged pipe

一级处理 primary treatment

引入管 service pipe, inlet pipe

用水量 water consumption

优质杂排水 high grade gray water

雨落水管 down pipe, leader

雨水斗 rain strainer

雨水口 gulley hole, gutter inlet

杂排水 gray water

中水 reclaimed water

中水 reclaimed water

中水设施 equipments and facilities of reclaimed water

中水系统 reclaimed water system

自用水量 water consumption in water works

综合生活用水 demand for domastic and public use

最小服务水头 minimum service head

建筑消防设计术语

安全出口 safety exit

安装间距 spacing

保护半径 monitoring radius

保护面积 monitoring area

报警区域 alarm zone

爆炸下限 lower explosion limit

闭式系统 close – type sprinkler system

边墙型扩展覆盖喷头 extended coverage sidewall sprinkler

标准喷头 standard sprinkler

不燃烧体 non – combustible component

重复启闭预作用系统 recycling preaction system

短立管 sprig – up

防护冷却水幕 drencher for cooling protection

防火分隔水幕 water curtain for fire compartment

防火分区 fire compartment

防火间距 fire separation distance

防火幕 fire curtain

防烟分区 smoke bay

防烟楼梯间 smoke proof staircase

封闭楼梯间 enclosed staircase

干式系统 dry pipe system

火灾备用照明 reserve lighting for fire risk

火灾疏散标志灯 marking lamp for fire evacuation

火灾疏收照明 lighting for fire evacuation

火灾疏散照明灯 light for fire evacuation

集中报警系统 remote alarm system

局部应用灭火系统 local application extinguishing system

控制中心报警系统 control center alarm system

快速响应喷头 fast response sprinkler

明火地点 open flame site

耐火极限 duration of fire resistance

耐火极限 fire resistance rating

难燃烧体 difficult – combustible component

泡沫—水雨淋系统 foam – water deluge system

配水干管 feed mains

配水管 cross mains

配水管道 system pipes

配水支管 branch lines

区域报警系统 local alarm system

全淹没灭火系统 total flooding extinguishing system

燃烧体 combustible component

湿式系统 wet pipe system

水幕系统 drencher systems

探测区域 detection zone

响应时间指数 response time index（RTI）

信号阀 signal valve

雨淋系统 deluge systemn

预作用系统 preaction system

早期抑制快速响应喷头 early suppression fast response sprinkler（ESFR）

自动喷水灭火系统 sprinkler systems

自动喷水—泡沫联用系统 combined sprinkler – foam

组合分配系统 combineddistribution systems

作用面积 area of sprinklers operation

建筑照明设计术语

安全照明 safety lighting

备用照明 stand – by lighting

不舒适眩光 discomfort glare

采光 daylighting

采光系数 daylight factor

采光系数标准值 standard value of daylight factor

参考平面 reference surface

灯具效率 luminaire efficiency

灯具遮光角 shielding angle of luminaire

发光强度 luminous intensity

反射比 reflectance

反射眩光 glare by reflection

分区一般照明 localied lighting

光幕反射 veiling reflection

光强分布 distribution of luminous intensity

光通量 luminous flux

光源的发光效能 luminous efficacy of a source

混合照明 mixed lighting

警卫照明 security lighting

局部照明 local lighting

亮度 luminance

亮度对比 luminance contrast

绿色照明 green lights

频闪效应 stroboscopic effect

日照标准 insolation standards

色温度 colour temperature

识别对象 recognized objective

视觉作业 visual task

疏散照明 escape lighting

特殊显色指数 special colour rendering index

统一眩光值 unified glare rating（UGB）

维持平均照度 maintained average illuminance

显色性 colour rendering

显色指数 colour rendering index

眩光 glare

眩光值 glare rating（GR）

一般显色指数 general colour rendering index

一般照明 general lighting

应急照明 emergency lighting

障碍照明 obstacle lighting

照度 illuminance

照度均匀度 uniformity ratio of illuminance

正常照明 normal lighting

直接眩光 direct glare

值班照明 on – duty lighting

作业面 working plane

采暖通风设计术语

变制冷剂流量多联分体式空调调节系统 variable refrigerant volume split air conditioning system

低温送风空气调节系统 cold air distribution system

地源热泵 ground – source pump

分区两管制水系统 zoning two – pipe water system

活动区 occupied zone

空气分布特性指标 air diffusion performance index

空气源热泵 air – source heat pump

湿球黑球温度指数 wet – bulack globe temperature index (WBGT)

水环境泵空气调节系统 water – loop heat pump air conditioning system

水源热泵 water – source heat pump

通风 ventilation

预计不满意者的百分数 predicted percentage of dissatisfied (PPD)

预计平均热感觉指数 predicted mean vote (PMV)

置换通风 displacement ventilation

建筑室内环境术语

超微粒子 ultrafine particle

洁净度 cleanliness

洁净工作区 clean working area

洁净工作台 clean bench

洁净区 clean zone

洁净室 clean room

粒径 partical size

粒径分布 particle size distribution

人身净化用室 room for cleaning human body
微粒子 macroparticle
物料净化用室 room for cleaning material
悬浮粒子 airborne particles

参 考 书 目

一、中文文献

[1] 包惠南,包昂. 中国文化汉英翻译 [M]. 北京:外文出版社,2004.

[2] 蔡基刚. 大学英语翻译教程 [M]. 上海:复旦大学出版社,2003.

[3] 陈安定. 英汉比较与翻译 [M]. 北京:中国对外翻译出版公司,1998.

[4] 陈福康. 中国译学理论史稿 [M]. 上海:上海外语教育出版社,1992.

[5] 陈宏薇. 汉英翻译基础 [M]. 上海:上海外语教育出版社,1998.

[6] 陈宏薇. 新实用汉英翻译教程 [M]. 武汉:湖北教育出版社,1996.

[7] 陈宏薇. 新编汉英翻译教程 [M]. 上海:上海外语教育出版社,2004.

[8] 陈新. 英汉文体翻译教程 [M]. 北京:北京大学出版社,2004.

[9] 程立,程建华. 英汉文化比较辞典 [M]. 长沙:湖南教育出版社,2000.

[10] 崔正勤. 英语比较结构 [M]. 济南:山东教育出版社,1986.

[11] 戴文进，章卫国．自动化专业英语［M］．武汉：武汉理工大学出版社，2001．

[12] 戴文进．科技英语翻译理论与技巧［M］．上海：上海外语教育出版社，2003．

[13] 丁树德．英汉汉英翻译教学综合指导［M］．天津：天津大学出版社，1996．

[14] 杜建慧等．翻译学概论［M］．北京：民族出版社，1998．

[15] 范武邱．实用科技英语翻译讲评［M］．北京：外文出版社，2001．

[16] 范仲英．实用翻译教程［M］．北京：外语教学与研究出版社，1994．

[17] 方梦之．翻译新论与实践［M］．青岛：青岛出版社，2002．

[18] 方梦之．译学辞典［M］．上海：上海外语教育出版社，2004．

[19] 费亚夫．英语否定结构的表达与翻译［M］．福州：福建人民出版社，1991．

[20] 傅似逸．英汉实用书信手册［M］．北京：北京大学出版社，1999．

[21] 傅晓玲等．英汉互译高级教程［M］．广州：中山大学出版社，2004．

[22] 耿伯华，戎林海．科技英语翻译 ABC（附录一），科技英语读本［M］．北京：国防工业出版社，2006．

[23] 耿秉钧．商用英文与国贸实务［M］．北京：世界图书出版公司，1999．

[24] 郭建中．当代美国翻译理论［M］．武汉：湖北教育出版社，2000．

[25] 郭著章，李庆生．英汉互译实用教程［M］．武汉：武汉大学出版社，1998．

[26] 胡庚申等. 国际商务合同起草与翻译 [M]. 北京：外文出版社，2001.

[27] 胡壮麟. 语篇的衔接与连贯 [M]. 上海：上海外语教育出版社，1994.

[28] 华先发. 英语的否定 [M]. 南宁：广西人民出版社，1985.

[29] 黄洋楼. 英汉互译实用技巧 [M]. 广州：华南理工大学出版社，2004.

[30] 黄忠康. 变译理论 [M]. 北京：中国对外翻译出版公司，2002.

[31] 贾德江. 英汉语对比研究与翻译 [M]. 长沙：国防科技大学出版社，2002.

[32] 宋天锡. 翻译新概念——汉互译实用教程（第三版）[M]. 北京：国防工业出版社，2005.

[33] 谭载喜. 西方翻译简史 [M]. 北京：商务印书馆，1991.

[34] 王福祯. 英语杏定句 [M]. 北京：外文出版社，2000.

[35] 王佩纶. 科技英语写作基础 [M]. 南京：东南大学出版社，1989.

[36] 王庆肇. 英语否定结构 [M]. 北京：商务印书馆，1982.

[37] 王佐良. 翻译：思考与试笔 [M]. 北京：外语教学与研究出版社，1989.

[38] 汪涛. 实用英汉互译技巧 [M]. 武汉：武汉大学出版社，2001.

[39] 吴森. 英汉比较与翻译 [M]. 北京：中国对外翻译出版公司，1988.

[40] 夏廷德. 翻译补偿研究 [M]. 武汉：湖北教育出版社，2006.

[41] 许建平. 英汉互译实践与技巧（第二版）[M]. 北京：清

华大学出版社，2004.

[42] 许钧等. 当代法国翻译理论 [M]. 南京：南京大学出版社，1998.

[43] 许明武. 新闻英语与翻译 [M]. 北京：中国对外翻译出版公司，2003.

[44] 许余龙. 对比语言学 [M]. 上海：上海外语教育出版社，2002.

[45] 严俊仁. 科技英语翻译技巧 [M]. 北京：国防工业出版，2000.

[46] 严俊仁. 汉英科技翻译 [M]. 北京：国防工业出版，2004.

[47] 袁昌明编. 英语比较结构 [M]. 北京：对外贸易教育出版社，1988.

[48] 臧仲伦. 中国翻译史话 [M]. 济南：山东教育出版社，1991.

[49] 曾昭智. 英语句子的否定形式 [M]. 武汉：武汉大学出版社，1995.

[50] 张健. 报刊新词英译纵横 [M]. 上海：上海科技教育出版社，2001.

[51] 张健. 新闻翻译教程 [M]. 上海：上海外语教育出版社，2008.

[52] 张今，张宁. 文学翻译原理 [M]. 北京：清华大学出版社，2005.

[53] 张梅岗. 科技英语修辞 [M]. 长沙：湖南科技出版社，1998.

[54] 张培基. 英汉翻译教程 [M]. 上海：上海外语教育出版社，1980.

[55] 张培基. 英译中国现代散文选 [M]. 上海：上海外语教

育出版社，1999.

[56] 张新红等．商务英语翻译［M］．北京：高等教育出版社，2003.

[57] 钟述孔．实用口译手册［M］．北京：中国对外翻译出版社公司，1999.

[58] 周方珠．翻译多元论［M］．北京：中国对外翻译出版公司，2004.

[59] 周静萍，李青林．实用英语应用文［M］．长沙：湖南出版社，1997.

[60] 周晓，周怡．现代英语广告［M］．上海：上海外语教育出版社，1998.

[61] 周志培．汉英对比与翻译中的转换［M］．上海：华东理工大学出版社，2003.

[62] 朱徽．汉英翻译教程［M］．重庆：重庆大学出版社，2004.

[63] 杨敏，纪爱梅．庄绎传．英汉翻译简明教程［M］．北京：外语教学与研究出版社，2002.

[64] 杨全红．英文广告口号的特点及其翻译［J］．上海科技翻译，1996（4）．

[65] 开晓予．英语形容词的理解与翻译［J］．安阳师范学院学报，2004（6）．

[66] 原传道．英语"信息型文本"翻译策略［J］．中国科技翻译，2005（3）．

[67] 曾立．试论广告文体英译策略［J］．株洲工学院学报，1996（3）．

[68] 曾利沙．论投资指南的汉英翻译原则［J］．国际经贸探索，2000（2）．

[69] 张健．英语对外报道并非逐字英译［J］．上海科技翻译，

2001 (4).

[70] 张铭. 浅述英语副词的汉译方法 [J]. 新疆职工大学学报, 1997 (3).

[71] 朱明炬. 专有名词翻译刍议 [J]. 武汉冶金科技大学学报 (社会科学版), 1999 (1).

[72] 朱益平. 论旅游翻译中文化差异的处理 [J]. 西北大学学报, 2005 (3).

[73] 贾文波. 汉英时文翻译 [M]. 北京: 中国对外翻译出版公司, 1999.

[74] 贾文波. 应用翻译功能论 [M]. 北京: 中国对外翻译出版公司, 2004.

[75] 居祖纯. 汉英语篇翻译 [M]. 北京: 清华大学出版社, 1998.

[76] 雷跃捷. 新闻理论 [M]. 北京: 北京广播学院出版社, 1997.

[77] 李长栓. 非文学翻译理论与实践 [M]. 北京: 中国对外翻译出版公司, 2003.

[78] 李克兴, 张新红. 法律文本与法律翻译 [M]. 北京: 中国对外翻译出版公司, 2005.

[79] 李明. 英汉互动翻译教程 [M]. 武汉: 武汉大学出版社, 2006.

[80] 李文革. 西方翻译理论流派研究 [M]. 北京: 中国社会科学出版社, 2004.

[81] 李学平, 蓝俊翔. 科技汉译英指南 [M]. 南宁: 广西人民出版社, 1985.

[82] 李运兴. 英汉语篇翻译 [M]. 北京: 清华大学出版社, 1998.

[83] 李运兴. 语篇翻译引论 [M]. 北京: 中国对外翻译出版

公司，2000.

[84] 李运兴．汉英翻译教程［M］．北京：新华出版社2006.

[85] 连淑能．英汉对比研究［M］．北京：高等教育出版社，1993.

[86] 廖七一．当代西方翻译理论探索［M］．南京：译林出版社，2000.

[87] 林相周．英语理解与翻译［M］．上海：上海外语教育出版社，1998.

[88] 林相周，周国珍．科技英语理解与翻译：对照与注释［M］．上海：上海科学技术文献出版社，1982.

[89] 刘宓庆．文体与翻译［M］．北京：中国对外翻译出版公司，1998.

[90] 刘宓庆．汉英对比与翻译［M］．南昌：江西教育出版社，1992.

[91] 刘其中．新闻翻译教程［M］．北京：中国人民大学出版社，2004.

[92] 卢红梅．华夏文化与汉英翻译［M］．武汉：武汉大学出版社，2006.

[93] 陆愚珠．英语外贸应用文［M］．北京：对外经济贸易大学出版社，1999.

[94] 鲁忠义，彭聃龄．语篇理解研究［M］．北京：北京语言大学出版社，2002.

[95] 罗新璋．翻译论集［M］．北京：商务印书馆，1984.

[96] 吕瑞昌．汉英翻译教程［M］．西安：陕西人民出版社，1983.

[97] 吕叔湘．中国人学英语［M］．北京：中国社会科学出版社，2005.

[98] 马会娟．商务英语翻译教程［M］．北京：中国商务出版

社，2004.

[99] 马祖毅. 中国翻译简史——五四以前部分 [M]. 北京：中国对外翻译出版公司，2007.

[100] 潘文国. 汉英语对比纲要 [M]. 北京：北京语言大学出版社，1997.

[101] 戚雨村. 现代语言学的特点和发展趋势 [M]. 上海：上海外语教育出版社，2000.

[102] 钱汝敏. 篇章语用学概论 [M]. 北京：外语教学与研究出版社，2001.

[103] 任学良. 汉英比较语法 [M]. 北京：中国社会科学出版社，1981.

[104] 戎林海. 跨越文化障碍——与英美人交往面面观 [M]. 南京：东南大学出版社，2005.

[105] 邵志洪. 汉英对比翻译导论 [M]. 上海：华东理工大学出版社，2005.

[106] 沈苏儒. 对外报道业务基础 [M]. 北京：今日中国出版社，1990.

[107] 沈苏儒. 关于中译英对外译品的质量问题 [A]. 中译英技巧文集 [C]. 北京：中国对外翻译出版公司.

[108] 司显柱. 汉译英教程 [M]. 上海：东华大学出版，2006.

二、外文文献

[109] Nida, Eugene A. 1964. Towards a Science of Translating.

[110] Nida, Eugene A. and Charles A. Taber. 1969. Theory and Practice of Translation.

[111] Nida, Eugene A. 1975, Language Structure and Translation.

[112] Nida, Eugene A. 1982. Translating Meaning.

[113] de Waard, Jan and Eugene A. Nida 1986. From one Language to Another.

［114］ Nida, Eugene A. 1989. Paradoxes of Translation.

［115］ Nida, Eugene A. 1990. Isomorphic Relations and Translational Equivalence.

［116］ Nida, Eugene A. 1991. Breakthroughs in Translation.

［117］ Nida, Eugene A. 1991. Translation: Possible and Impossible.

［118］ Nida, Eugene A. 1991. Language as an Efficient Code.

［119］ Savory, Theodore. 1968. The Art of Translation.

［120］ A Dictionary of American Idioms.

［121］ Oxford Dictionary of Current Idiomatic English.

［122］ Steiner, George. After Babel: Aspects of language and Translation (3rd ed.) ［M］. London Oxford, 1998. Shuttleworth, Mark & Moira Cowie. Dictionary of Translation Studies ［Z］. Shanghai. Shanghai Foreign Language Education Press, 2004.

［123］ Snell-Hornby, Mary. Translation Studies: An Integrated Approach ［M］. Shanghai Shanghai Foreign Language Education Press, 2001.

［124］ Tanaka, Keilo. Advertising Language ［M］. London: Routledge. 1994.

［125］ Toury, Gideon. In Search of a Theory Translation ［M］. Tel Aviv: The Porter Institute for Poetics and Semiotics, 1980.

［126］ Venuti, Lawrence. The Translation's Invisibility. London & New York: Routledge 1995.

［127］ Mukul G. Asher: Extending social security coverage in Asia Pacific: A review ofgood practices and lessons Learnt ［J］. International Social Security Association – Geneva. 2009.

［128］ Donghyun Park and Gernma B. Estrada: Developing Asia's Pension Systems and Old-Age Income Support ［J］. ADBI Working Paper Series, 2012, 4.